# MCQs in Physiology

# MCQs in Physiology

## (With Explanatory Answers)

**Jayant J Makwana**
MD (Physiology)
Associate Professor
LTMMC, Sion
Mumbai

**Surendra S Wadikar**
MD (Physiology)
Lecturer, TNMC
Mumbai

**JAYPEE BROTHERS**
**MEDICAL PUBLISHERS (P) LTD**
New Delhi

*Published by*

Jitendar P Vij
**Jaypee Brothers Medical Publishers (P) Ltd**
EMCA House, 23/23B Ansari Road, Daryaganj,
**New Delhi** 110 002,
Phones: +91-11-23272143, 23272703, 23282021,
+91-11-23245672, Fax: 011-23276490, 23245683
e-mail: jaypee@jaypeebrothers.com
Visit our website: www.jaypeebrothers.com

*Branches*

- 2/B Akruti Society, Jodhpur Gam Road, Satellite **Ahmedabad** 380 015, Phone: +91-079-30988717, 26926233 e-mail: jpamdvd@rediffmail.com
- 202 Batavia Chambers, 8 Kumara Krupa Road, Kumara Park East, **Bangalore** 560 001, Phones: +91-80-22285971, 22382956, +91-80-30614073, Tele Fax: +91-80-22281761 e-mail: jaypeemedpubbgl@eth.net
- 282 IIIrd Floor, Khaleel Shirazi Estate, Fountain Plaza Pantheon Road, **Chennai** 600 008, Phones: +91-44-28262665, 28269897, Fax: +91-44-28262331 e-mail: jpchen@eth.net
- 4-2-1067/1-3, Ist Floor, Balaji Building, Ramkote Cross Road, **Hyderabad** 500 095, Phones: +91-40-55610020, 24758498 Fax: +91-40-24758499 e-mail: jpmedpub@rediffmail.com
- No. 41/3098, B & B1, Kuruvi Building, St. Vincent Road, **Kochi** 682 018, Kerala. Phones: +91-0484-4036109
- 1A Indian Mirror Street, Wellington Square, **Kolkata** 700 013, Phones: +91-33-22456075, +91-33-22451926 Fax: +91-33-22456075 e-mail: jpbcal@cal.vsnl.net.in
- 106 Amit Industrial Estate, 61 Dr SS Rao Road, Near MGM Hospital Parel, **Mumbai** 400 012, Phones: +91-22-24124863, +91-22-24104532, +91-22-30926896 Fax: +91-22-24160828 e-mail: jpmedpub@bom7.vsnl.net.in
- "KAMALPUSHPA",38 Reshimbag, Opp. Mohota Science College, Umred Road **Nagpur** 440 009, Phones: 0712-3945220, 2704275, Fax: +91-0712-2704275 e-mail: jpmednagpur@rediffmail.com

***MCQs in Physiology***



First Edition : 2005
Reprint : 2006

ISBN 81-8061-428-X

*Typeset at* JPBMP typesetting unit
*Printed at* Gopsons Papers Ltd, A-14, Sector 60, Noida

*to*

*My Dearest*

Nephew Nikhil

and

Niece Priyanka

# Preface

In this ever increasingly competitive exam, the MCQs are gaining more and more important role. Through this book we have tried to cover the requirements of Medical, Dental, Nursing, Occupational Therapy and Physiotherapy students. The book also will help to cater to the requirement of the students appearing for various competitive exams. The students usually read the books on MCQs, which gives the correct answer, but students are not sure why the particular choice is the right one. Through this book we have tried to rationalize the correct choice, which would make students understanding about the subject quite clear.

**Jayant J Makwana**
**Surendra S Wadikar**

# Acknowledgements

We thank our family members without whose moral support it would not have been possible to bring out this book. We also would like to thank Mr CS Gawde (Mumbai Region representative of Jaypee Brothers), who was always there to solve any queries about publishing the book. Lastly our sincere thanks to GOD.

# Contents

# 1

# Haematology

**1. The stage of erythropoiesis during which haemoglobin makes it appearance is**

**A.** CFU – ε
**B.** Proerythroblast
**C.** Intermediate normoblast
**D.** Reticulocyte

**Ans. 1. (C)**

It is the stage where staining capability of haemoglobin becomes polychromatophilic : CFU – ε - it is a committed stem cell for the production of erythrocytes. Proerythroblast – it is the first identifiable cell of erythroblastic series reticulocyte – it is a stage after the extrusion of nuclei from late normoblasts.

**2. The haemoglobin A has**

**A.** 2 α and 2 β chains
**B.** 2 α and 2δ chains
**C.** 2 δ and 2 β chains
**D.** 2 γ and 2 α chains

**Ans. 2. (A)**

Hemoglobin A is the form of hemoglobin present in adults. D-this chains are present in fetal hemoglobin [Hemoglobin F]. B-this chains are present in varient of adult hemoglobin [Hemoglobin A2]. C-this form does not exist].

**3. The maximum amount of haemoglobin that can be concentrated in RBC is**

**A.** 16%
**B.** 34%
**C.** 29 gm%
**D.** 14.5 gm%

**Ans. 3. (B)**

D–that is the average amount of hemoglobin present in the 100 ml of blood.

**4. Each haemoglobin molecule can transport**

**A.** 8 molecules of oxygen
**B.** 4 molecules of oxygen

C. 2 molecules of oxygen
D. 4 atoms of oxygen

**Ans. 4. (B)**

Because of 4 heme group [one for each chain], there are 4 iron atoms in each hemoglobin molecule, which can bind with 4 molecules of oxygen [8 atoms of oxygen].

**5. The affinity of Hb for oxygen depends on**

**A.** Type of chains in Hb
**B.** Hb conc. in blood
**C.** Conc. of oxygen in blood
**D.** Erythropoietin conc. in blood

**Ans. 5. (A)**

The affinity of Hb for $O_2$ depends on the type of chains. With normal level of Hb having altered chain will affect the $O_2$ carriage. Erythropoietin affects the rate of synthesis of RBCs.

**6. Time taken for the formation of RBC from proerythroblast is**

**A.** 1 month **B.** 120 days
**C.** 9-10 days **D.** 3 weeks

**Ans. 6. (C)**

120 days is the normal life span of RBCs.

**7. RBCs can form small ATP from glucose which is metabolized in**

**A.** Nucleus
**B.** Mitochondria
**C.** Cytoplasm
**D.** Endoplasmic reticulum

**Ans. 7. (C)**

RBCs are non-nucleated and produce ATP through glucolytic pathways, enzymes for which are present in cytoplasm.

**8. The NADPH formed in RBC helps in all of the following EXCEPT**

**A.** Maintaining cell membrane pliability
**B.** Maintenance of ion transport across the membrane
**C.** To keep iron of Hb in ferric form
**D.** To prevent proteins in RBCs from being oxidised (ferric form causes formation of meth-Hb that will not carry $O_2$)

**Ans. 8. (C)**

Ferric form causes formation of meth-Hb that will not carry $O_2$. NADH methemoglobin reductase system helps to keep iron in ferrous form.

**9. Hemoglobin constitutes to about**

**A.** 0.1% of the total body iron
**B.** 65% of the total body iron
**C.** 1% of the total body iron
**D.** 15-30% of the total body iron

**Ans. 9. (B)**

0.1% in the form of transferritin, 1% in the form of various heme compound, 15-30% in the RES and liver parenchyma as ferritin.

**10. Vitamin $B_{12}$ and folic acid helps in maturation of RBC because they are essential for the synthesis of**

**A.** RNA **B.** DNA
**C.** Golgi complex **D.** Endoplasmic reticulum

**Ans. 10. B**

**11. Which of the following statement about macrocyte (because of vitamin $B_{12}$ and folic acid deficiency) is true**

**A.** They are biconcave in shape
**B.** Their $T_{1/2}$ is same as that of normal RBCs
**C.** After entering circulation it has normal $O_2$ carrying capacity
**D.** Its $O_2$ carrying capacity is less

**Ans. 11. (C)**

Macrocytes have abnormal shape. The increased membrane fragility reduces there half-life but doesn't affect the $O_2$ carrying capacity.

**12. Fractional precipitation means**

**A.** Separation of plasma proteins by passing current through the solution
**B.** Separation of plasma proteins by using different concentrations of a salt
**C.** Separation of plasma proteins by using ultracentrifuge
**D.** Method to study ionization of proteins

**Ans. 12. (B)**
Electrophoresis, ultracentrifugation, plasma-pheresis

**13. Electrophoresis means**
- **A.** Separation of plasma proteins by passing current through the solution
- **B.** Separation of plasma proteins by using different concentrations of a salt
- **C.** Separation of plasma proteins by using ultra-centrifuge
- **D.** Method to study ionization of proteins

**Ans. 13. (A)**
Fractional precipitation, ultracentrifugation, plasma-pheresis.

**14. Plasmapheresis means**
- **A.** Separation of plasma proteins by passing current through the solution
- **B.** Separation of plasma proteins by using different concentrations of a salt
- **C.** Separation of plasma proteins by using ultra-centrifuge
- **D.** Method to study ionization of proteins

**Ans. 14. (D)**
Electrophoresis, Fractional precipitation, ultracentri-fugation.

**15. Normal percentage of neutrophil in the adult human is**
- **A.** 50 - 70% **B.** 20 - 40 %
- **C.** 02 - 08% **D.** 00 - 01%

**Ans. 15. (A)**

**16. Normal percentage of monocyte in the adult human is**
- **A.** 50 - 70% **B.** 20 - 40 %
- **C.** 02 - 08% **D.** 00 - 01%

**Ans. 16. (C)**

**17. Normal percentage of eosinophil in the adult human is**
- **A.** 50 - 70% **B.** 20 - 40 %
- **C.** 01 - 04% **D.** 00 - 01%

**Ans. 17. (C)**

**18. Normal percentage of lymphocyte in the adult human is**

**A.** 10 - 20% **B.** 20 - 30 %
**C.** 02 - 08% **D.** 00 - 01%

**Ans. 18. (B)**

**19. Neutrophil releases**

**A.** Histamine
**B.** Myeloperoxidase
**C.** Interleukin I
**D.** Proteolytic enzymes

**Ans. 19. (B)**

19-21 histamine is released by basophils and mast cells, interleukin is released by macrophages, proteolytic enzymes are released by lysosomes.

**20. Macrophage releases**

**A.** Histamine
**B.** Myeloperoxidase
**C.** Interleukin I
**D.** Proteolytic enzymes

**Ans. 20. (C)**

**21. Basophil releases**

**A.** Histamine
**B.** Myeloperoxidase
**C.** Interleukin I
**D.** Proteolytic enzymes

**Ans. 21. (A)**

**22. Normocytic normochromic anaemia is seen in**

**A.** The deficiency of vitamin $B_{12}$
**B.** The deficiency of folic acid
**C.** The deficiency of iron
**D.** Acute haemorrhage

**Ans. 22. (D)**

Acute loss of blood does not lead to any specific nutrient loss. The basic change is in the number of RBCs in the blood which is less and hence the anaemia.

**23. Microcytic hypochromic anaemia is seen in the deficiency of**

**A.** Cyanocobalamin **B.** Folic acid
**C.** Iron **D.** Intrinsic factor

**Ans. 23. (C)**

Deficiency of all other factors lead to macrocytic anaemia.

**24. Macrocytic hypochromic anaemia is seen in the deficiency of**

A. Cyanocobalamin B. Riboflavin
C. Iron D. Vitamin C

**Ans. 24. (A)**

Iron deficiency leads to microcytic hypochromic anaemia. Vitamin C helps in absorption of ion.

**25. Person with acute haemorrhage should ideally be transfused with**

A. Plasma B. Whole blood
C. Isotonic saline D. Platelet

**Ans. 25. (B)**

Plasma is an ideal fluid in case of burns, isotonic saline in case of fluid loss and platelet for the hemorrhagic disorder.

**26. Erythropoiesis is inhibited by**

A. Polycythemia
B. Anaemia
C. Hypoxia
D. Erythropoietin

**Ans. 26. (A)**

All the other factors stimulate erythropoiesis. The stimulus for erythropoietin secretion is hypoxia.

**27. Acute dehydration in a person can be corrected by transfusing**

A. Plasma B. Whole blood
C. Isotonic saline D. Platelet

**Ans. 27. (C)**

Plasma is an ideal fluid in case of burns whole blood in case of hemorrhage and platelet for the hemorrhagic disorder.

**28. Erythrocyte sedimentation rate**

A. Is less in inflammation
B. Varies with the rate of rouleaux formation
C. Increases in anisocytosis
D. Increases in polycythemia

**Ans. 28. (B)**

Inflammatory process is associated with increase in ESR, anisocytosis decreases the rate of rouleaux and hence decreases the ESR, polycythemia decreases ESR because of repelling force and less space for sedimentation.

**29. Haemophilia is**

**A.** Common in females
**B.** Due to deficiency of factor VIII
**C.** An autoimmune disorder
**D.** Associated with prolonged bleeding time

**Ans. 29. (B)**

Females usually acts as a carrier, is associated with prolonged clotting time.

**30. Which of the following helps in humoral immunity**

**A.** Neutrophil **B.** T lymphocytes
**C.** Immunoglobulins **D.** Interferon

**Ans. 30. (B)**

Neutrophil forms the second line of defence and helps in phagocytosis, T lymphocytes helps in cell mediated immunity, interferon causes fever.

**31. Platelets help in clotting by release of**

**A.** Calcium **B.** Platelet factor 3
**C.** Contact factor **D.** Thrombopoietin

**Ans. 31. (B)**

**32. Person suffering from severe burns is transfused with**

**A.** Plasma **B.** Whole blood
**C.** Isotonic saline **D.** Platelet

**Ans. 32. (A)**

The basic pathology in burns is a loss of plasma isotonic saline in case of fluid loss and platelet for the hemorrhagic disorder, whole blood for the blood loss.

**33. Most of the erythropoietin is secreted by**

**A.** Kidney **B.** Liver
**C.** Spleen **D.** Lymph node

**Ans. 33. (A)**

About 90% of the erythropoietin is secreted by kidney and only 10% by liver, spleen and lymph node does not play any role in the secretion.

**34. After the age of 20 years the chief site for RBCs production is**

**A.** Proximal portion of femur
**B.** Liver
**C.** Lymph node
**D.** Membranous bones

**Ans. 34. (D)**

Lymph node and liver produces RBCs in the intrauterine life, femur forms RBCs till the age of 15–20.

**35. All of the following statements about erythroblastosis foetalis are true EXCEPT**

**A.** Is associated with jaundice
**B.** Is a hemolytic disease
**C.** Occurs in Rh negative babies
**D.** Can be prevented by injection of anti D to mother

**Ans. 35. (C)**

Erythroblastosis foetalis results when Rh –ve mother bears an Rh +ve child. It is an agglutination reaction between D antigen of a baby and anti D antibody of mother.

**36. True haematocrit is**

**A.** 96% of the measured hematocrit
**B.** 91% of the measured hematocrit
**C.** 87% of the measured hematocrit
**D.** Same as measured hematocrit

**Ans. 36. (A)**

Hematocrit is the RBCs portion of the blood, which is measured by centrifuging the blood, but after this also some amount of plasma remains trapped in between the cell which comes to about 4%.

**37. Average blood volume of a normal adult is**

**A.** 60% of total body weight
**B.** 8% of total body weight
**C.** 40% of total body weight
**D.** 20% of total body weight

**Ans. 37. (B)**

Total body water averages 60%, ICF averages 40% and ECF averages 20%.

**38. The plasma lost during rapid blood loss is replaced by body within**

**A.** 1–3 days **B.** 1–3 hours
**C.** 3–6 weeks **D.** 1–2 months

**Ans. 38. (A)**

**39. After a single bout of haemorrhage the RBCs returns to the normal values within**

**A.** 3–6 hours **B.** 3–6 days
**C.** 3–6 weeks **D.** 3–6 months

Ans. 39. (C)

**40. The term bone marrow aplasia refers to**

**A.** Increase functioning of bone marrow
**B.** Lack of functioning of bone marrow
**C.** Formation of red marrow from yellow marrow
**D.** Increased cellularity of marrow

**Ans. 40. (B)**

**41. Megaloblastic anaemia can be the result of deficiency of all of the following EXCEPT**

**A.** Extrinsic factor
**B.** Folic acid
**C.** Iron
**D.** Intrinsic factor

**Ans. 41. (C)**

Iron deficiency causes the formation of the RBCs which are smaller in size percent lesser in Hemoglobin concentration.

**42. Hemolytic anaemia is observed in all of the following EXCEPT**

**A.** Sickle cell anaemia
**B.** Mismatched blood transfusion
**C.** Erythroblastosis foetalis
**D.** Folic acid deficiency

**Ans. 42. (D)**

Hemolytic anaemia results from the hemolysis of the RBCs as in antigen antibody reaction (B and C), or abnormal hemoglobin as in sickle shape anaemia, folic acid deficiency leads to nutritional anaemia.

**43. Which of the following statement regarding bilirubin is true**

**A.** It is a steroid pigment
**B.** It is carried to liver in a conjugate form
**C.** It contains iron
**D.** Its normal level is 0.2 - 0.8 mg/dl

**Ans. 43. (D)**

It is a bile pigment formed from porphyrins, which is carried to liver in the bound form with plasma proteins and it gets conjugated in liver and is devoid of iron.

**44. The life span of monocyte in the tissue is**

**A.** 4–5 days
**B.** 4–5 hours

**C.** Till destroyed while phagocytosis
**D.** 4–5 weeks

**Ans. 44. (C)**

Monocytes circulates in the blood for few hours and then enters the tissue and becomes tissue macrophage, in this form it can live till destroyed when it participates in the immune reaction.

**45. The life span of neutrophil in the blood is**
**A.** 4–8 hours in blood
**B.** 18–20 hours in blood
**C.** 120 days
**D.** 7 days

**Ans. 45. (A)**

**46. The life span of RBC in the blood is**
**A.** 120 hours **B.** 12 days
**C.** 120 days **D.** 120 weeks

**Ans. 46. (C)**

**47. Which of the following is the first line of defense after the inflammation sets in**
**A.** Invasion of injured area by neutrophil
**B.** Second macrophage invasion
**C.** Tissue macrophage
**D.** Increased WBC production

**Ans. 47. (C)**

Any route through which antigens enters the body it first comes in the contact with the the tissue macrophage.

**48. The eosinophil kills the worm by all of the following mechanism EXCEPT**
**A.** By releasing hydrolytic enzymes
**B.** By phagocytosis
**C.** By releasing major basic proteins
**D.** By releasing highly reactive form of oxygen

**Ans. 48. (B)**

A worm is bigger than the eosinophil and hence it can not be phagocytosed by eosinophil.

**49. Which of the following statement about innate immunity is FALSE**
**A.** It is inborn
**B.** It is disease specific
**C.** It is species specific
**D.** N K lymphocyte helps in innate immunity

**Ans. 49. (B)**

Innate immunity is an immunity which one acquires as belonging to the particular species and is general.

**50. The IgG antibodies**

**A.** Are found in bronchial secretions

**B.** Are concerned with hypersensitive states

**C.** Appears during primary response

**D.** Appears during secondary immune response

**Ans. 50. (D)**

Ig A is present in the tissue secretion, Ig E is associated with the hypersensitive reactions and IgM plays an important role in primary immune response.

**51. The first clotting factor to be activated during extrinsic mechanism of blood coagulation is**

**A.** Factor XII **B.** Factor VII

**C.** Factor XI **D.** Factor I

**Ans. 51. (B)**

Factor XII is the first factor to be activated in intrinsic mechanism, Factor I is the final product.

**52. Frequency of people with blood group A in Western population is**

**A.** 9% **B.** 41%

**C.** 47% **D.** 25%

**Ans. 52. (B)**

**53. The site of origin of T lymphocyte is**

**A.** Lymph node **B.** Spleen

**C.** Bone marrow **D.** Thymus

**Ans. 53. (C)**

Lymphocytes originate in bone marrow and they are further processed in thymus and are stored in lymph node and spleen.

**54. The preprocessing of T lymphocyte in thymus usually occurs**

**A.** Just after or before birth

**B.** At puberty

**C.** When exposed to particular antigen

**D.** None of the above

**Ans. 54. (A)**

**55. The variable portion of the antibody is responsible for**

A. Diffusibility of antibody
B. Attachment of antibody to complement complex
C. Attachment of antibody to the tissue
D. Specificity of antibody for antigen

**Ans. 55. (D)**

All other properties depend on the constant portion.

**56. Which class of the antibody has the highest concentration in plasma**

A. IgG B. IgM
C. IgA D. IgE

**Ans. 56. (A)**

IgG – 75% (1000 mg/dl), IgM (120 mg/dl,
IgA 15 – 20% (200 mg/dl), IgE (0.05 mg/dl)

**57. Which class of antibody plays a important role during a primary immune response**

A. IgM B. IgG
C. IgA D. IgE

**Ans. 57. (A)**

IgG is important during secondary immune response.

**58. A person with a blood group O will have**

A. Antigen A and B on RBC
B. Antibody A and B on RBC
C. Antigen A on RBC
D. A and B antibody in plasma

**Ans. 58. (D)**

**59. Frequency of people with blood group O in Indian population is**

A. 10% B. 45%
C. 25% D. 35%

**Ans. 59. (B)**

**60. Frequency of people with blood group O in Western population is**

A. 9% B. 40%
C. 47% D. 25%

**Ans. 60. (C)**

**61. Frequency of people with blood group A in Indian population is**

A. 41% B. 9%
C. 47% D. 25%

**Ans. 61. (D)**

**62. The IgM antibodies**

A. Are found in bronchial secretions
B. Appears during primary response
C. Concerned with hypersensitive states
D. Appears during secondary immune response

**Ans. 62. (B)**

IgA is found in bronchial secretions, IgE is concerned with the hypersensitive states and IgG is concerned with secondary immune response.

**63. Frequency of people with blood group B in Western population is**

A. 41% B. 50%
C. 47% D. 9%

**Ans. 63. (D)**

**64. Frequency of people with blood group B in Indian population is**

A. 9% B. 45%
C. 25% D. 35%

**Ans. 64. (C)**

**65. Frequency of people with blood group AB in Western population is**

A. 30% B. 3% C. 45% D. 25%

**Ans. 65. (B)**

**66. Frequency of people with blood group AB in Indian population is**

A. 5-9% B. 40%
C. 47% D. 25%

**Ans. 66. (A)**

**67. Which of the following is true about stored blood**

A. Decrease in intracellular $Na^+$
B. Increase in intracellular $Na^+$
C. Decrease in plasma $K^+$
D. Hypercalcemia in plasma

**Ans. 67. (B)**

During the storage there is decrease in the energy providing substrate and hence decrease in the activity of ATPase pump, this leads to increase transport of sodium and calcium in the cell and potassium out of the cell.

**68. Major cross matching is tested between**
**A.** Donor's cells and recipient's plasma
**B.** Donor's plasma and recipient's cells
**C.** Donor's and recipient's plasma
**D.** Donor's and recipient's cells

**Ans. 68. (A)**

B is an example of minor cross matching as donor's plasma gets diluted in 5 litres receipient's of blood.

**69. Platelets are formed by fragmentation of**
**A.** Megaloblasts **B.** Myeloblasts
**C.** Megakaryocytes **D.** Microcytes

**Ans. 69. (C)**

**70. The secondary immune response following the exposure to an antigen is**
**A.** More potent and long lasting
**B.** Weak and long lasting
**C.** More potent and short lived
**D.** Weak and short lived

**Ans. 70. (A)**

Carried out by IgG and there ia a presence of memory cell which helps to detect an antigen and bring about the response.

**71. The titre of antibody (A and B agglutinins) which is zero at the time of birth reaches its peak value at the age of**
**A.** 2–8 months **B.** 30 years
**C.** At puberty **D.** 8–10 years

**Ans. 71. (D)**

At the time of birth because of low level of exposure the titre of antibodies A and B in the body is negligible and with subsequent exposure the titre rises.

**72. The 1st mechanism to come into play to reduce the blood loss when the vessel is injured is**
**A.** Formation of platelet plug
**B.** Clot formation
**C.** Vasospasm
**D.** Clot retraction

**Ans. 72. (C)**

Whenever there is an injury there is a release of local humoral factors which brings about vasospasm along with the neural mechanism because of pain.

**73. The IgE antibodies**

**A.** Are found in bronchial secretions
**B.** Are concerned with hypersensitive states
**C.** Appears during primary response
**D.** Appears during secondary immune response

**Ans. 73. (B)**

IgA is found in bronchial secretions, IgM appears during primary immune response hypersensitive states and IgG is concerned with secondary immune response.

**74. The first clotting factor to be activated during intrinsic mechanism of blood coagulation is**

**A.** Factor X **B.** Factor II
**C.** Factor IX **D.** Factor XII

**Ans. 74. (D)**

Intrinsic mechanism is activated by injury to the vessel wall or changes in blood, which causes the activation of factor XII.

**75. The clotting of blood in normal intact vascular segment is prevented by all the following EXCEPT**

**A.** Smooth endothelium
**B.** Absence of glycocalyx on endothelium
**C.** Presence of thrombomodulin
**D.** Continuous blood flow

**Ans. 75. (B)**

A glycocalyx covering on the endothelium prevents the interaction between clotting factors and the endothelium, also rough endothelium is a base for the platelets and other clotting factors to attach to.

**76. Vitamin K is essential for the formation of**

**A.** Fibrinogen **B.** Factor XII
**C.** Factor VII **D.** Factor V

**Ans. 76. (C)**

**77. Haemophilia A (Classical Haemophilia) is caused by deficiency of**

**A.** Factor I **B.** Factor IX
**C.** Factor VIII **D.** Calcium

**Ans. 77. (C)**

**78. Which of the following statement about haemophilia is NOT TRUE**

**A.** It is genetically transmitted as a dominant trait
**B.** Clotting time is prolonged
**C.** Woman will act as a carrier
**D.** It is genetically transmitted as a recessive trait

**Ans. 78. (A)**

Females usually are carriers and one X in them is normal.

**79. Which of the following statement about thrombocytopenia is FALSE**

**A.** Bleeding is from small vessels
**B.** It is caused by platelet deficiency
**C.** Most common cause is idiopathic
**D.** Bleeding is because of deficient clot formation

**Ans. 79. (D)**

Deficiency is of platelet plug formation.

**80. Which of the following plays the more important role in determination of colloidal osmotic pressure**

**A.** Globulin **B.** Fibrinogen
**C.** Prothrombin **D.** Albumin

**Ans. 80. (D)**

Globulin is mostly a carrier protein and also helps in immunity, prothrombin and fibrinogen helps in clotting.

**81. Which of the exhibits fastest movement on electrophoresis**

**A.** Albumin **B.** $\gamma$ globulin
**C.** $\beta$ globulin **D.** $\alpha$ globulin

**Ans. 81. (A)**

Followed by $\alpha$, $\beta$, $\gamma$ globulin.

**82. Which of the following is not synthesized in liver**

**A.** Prothrombin **B.** Fibrinogen
**C.** Albumin **D.** Immunoglobulin

**Ans. 82. (D)**

Synthesized in the lymphoid tissue.

**83. Which of the plasma protein is first to be regenerated**

**A.** Albumin **B.** Globulin
**C.** Fibrinogen **D.** Prothrombin

**Ans. 83. (C)**

**84. The total quantity of iron in the body averages**

**A.** 4–5 gm **B.** 6–9 gm

**C.** 14–16 gm **D.** 10–12 gm

**Ans. 84. (A)**

**85. RES and liver has about constitutes to about**

**A.** 0.1% of the total body iron

**B.** 65% of the total body iron

**C.** 1% of the total body iron

**D.** 15 - 30 % of the total body iron

**Ans. 85. (D)**

**86. The intermediatory role of antibody in binding of antigen to the phagocytic cell is called**

**A.** Agglutination

**B.** Rouleaux formation

**C.** Opsonization

**D.** Neutralization

**Ans. 86. (C)**

Adherence of antigen to is agglutination antibody while converting the harmful viruses to nonvirulent strain by complement is neutralization, rouleaux formation is just settling down of RBCs.

# 2

# Haemodynamics

1. **The velocity of blood flow is**
   **A.** Higher in capillaries than in arterioles
   **B.** Higher in veins than in venules
   **C.** Higher in veins than arteries
   **D.** Reduced in constricted area of the blood vessel

**Ans. 1. (B)**

The velocity of the blood flow is inversely proportional to the cross sectional area of the vessels. The cross sectional area of veins is less than that of the venules.

2. **Increase in the radius of the resistance vessel is associated with**
   **A.** Increase in systolic blood pressure
   **B.** Increase in diastolic blood pressure
   **C.** Increase in viscosity of blood
   **D.** Increase in capillary blood flow

**Ans. 2. (D)**

Radius of the vessel determines the diastolic pressure. Increase in the radius decreases the diastolic blood pressure and by decreasing the resistance it increases the blood flow.

3. **The pulse pressure wave in systemic arteries is initiated by**
   **A.** Ejection of the blood into the aorta
   **B.** Has same velocity as the arterial blood flow
   **C.** Faster in compliant vessel than noncompliant vessel
   **D.** Propagated at the rate of 0.7m/s

**Ans. 3. (A)**

The velocity of the arterial blood flow is about 1/10th of the pulse wave velocity. The pressure pulse propagates at about 7m/sec, the less compliant vessel propagates the wave at higher velocity.

4. **The pulse pressure is**
   **A.** 1/3rd of the difference between systolic and diastolic blood pressure
   **B.** Highest in capillaries
   **C.** Increases with decrease in the compliance of the vessel
   **D.** Independent of age

**Ans. 4. (C)**

The pulse pressure is difference between systolic and diastolic and more compliant the vessel, less will be the rise in the systolic pressure, with the advancing as vessel becomes less compliant, capillary pressure is non-pulsatile.

5. **The maximum resistance to the blood flow is offered by**
   **A.** Capillaries **B.** Arterioles
   **C.** Aorta **D.** Veins

**Ans. 5. (B)**

In the arterioles lumen to wall thickness ratio is less and also the wall has more muscular coating, capillaries are made up of single endothelial lining, aorta has more of elastic fibres.

6. **The wave in the aorta travels at the speed of**
   **A.** 3-5 m/s **B.** 7-10 m/s
   **C.** 15-35 m/s **D.** 0.5 mm/s

**Ans. 6. (A)**

Greater the compliance of the vessel, slower is the velocity and hence velocity in the aorta is less than large and small arteries.

7. **The metabolic theory for autoregulation proposes that**
   **A.** Local metabolites relax the precapillary sphincter
   **B.** Metabolism leads to the formation of the vasoconstrictor substance
   **C.** Adenosine is imp. In regulation of blood flow
   **D.** Stretch of vascular smooth muscle causes vasodilatation

**Ans. 7. (C)**

8. **Under normal physiological condition the resistance to blood flow is determined by**
   **A.** Length of the blood vessel
   **B.** Viscosity of blood

C. Pressure in the vessel
D. Radius of the vessel

**Ans. 8. (D)**

Doubling the pressure can double the flow, but doubling the radius increases the flow 16 times $R = 1/r^4$, viscosity and length usually does not change under normal circumstances.

**9. The mean arterial pressure in the upright posture**
A. At the level of the heart is approximately 60 mm of mercury
B. In the large artery in the brain less than 93 mm of mercury
C. In the large artery in the brain less than 93 mm of mercury
D. Does not differ from the reclining position

**Ans. 9. (B)**

The gravity has a significant effect on the arterial pressure and hence pressure in the vessel above the level of heart (where it is about 100 mmHg) decreases.

**10. Increase in the viscosity of the blood results in increase in the**
A. Mean arterial BP
B. Central venous pressure
C. Flow of the blood
D. Radius of the resistance vessels

**Ans. 10. (A)**

Viscosity is one of the determinant of the resistance and so of the diastolic pressure.

**11. All of the following statements regarding cerebral blood flow are true EXCEPT**
A. $O_2$ consumption of white matter is very high
B. Blood flow through the brain remains constant
C. Cranial circulation time is very short
D. Is dependent on effective perfusion pressure and cerebral vascular resistance

**Ans. 11. (A)**

It is the gray matter which has the high consumption rate of oxygen, out of total of 50 ml of $O_2$ consumption gray matter uses 40–45 ml though it constitutes only 1% of the total body weight.

**12. Volume flow through a rigid tube**

**A.** Doubles when the pressure difference between two ends of the tube is doubled

**B.** Increases 8 times when the radius is doubled

**C.** Is unchanged by doubling the length of the tube

**D.** Decreases with decrease in the viscosity

**Ans. 12. (A)**

Flow increases 16 times when the radius is doubled, flow will increase with the decrease in the viscosity, flow decreases by a factor of 2 when the length is doubled.

**13. When blood passes through a small blood vessel at moderate velocity**

**A.** A flow along the wall is faster than at the centre

**B.** The speed of flow is determined by the law of laplace

**C.** Turbulence occurs

**D.** RBCs are concentrated towards the middle of the stream

**Ans. 13. (D)**

Maximal flow velocity occurs in the center of the stream where the RBCs tend to concentrate. The law of laplace relates to forces in the wall of the vessel rather than blood flow velocity.

**14. According to Bernoulli's principle**

**A.** Total energy of streamline flow is constant

**B.** Total energy is the sum of potential and 1/2 kinetic

**C.** Lateral pressure distending a vessel is increased by constriction

**D.** There is greater lateral pressure with greater velocity of flowing blood

**Ans. 14. (A)**

Bernoulli's principles states that the total energy in fluids with streamlined flow is constant and is equal to the sum of its potential (pressure) and kinetic energy (flow velocity).

**15. Critical closing pressure of a blood vessel**

**A.** Decreases during adrenergic $\alpha_1$ – receptor activation

**B.** Decreases if extravascular pressure decreases

**C.** Increases during adrenergic $\beta_2$ – receptor stimulation

**D.** Does not change if the blood vessel become compliant

**Ans. 15. (B)**

Critical closing pressure increases if the vessel become less compliant and decreases from smooth muscle relaxation because of $\beta_2$ – receptor stimulation.

**16. All of the following statement about blood flow velocity is true EXCEPT**

**A.** It is not a major determinant of pulse pressure
**B.** It is greatest in the capillaries
**C.** It increases with the decrease in the vessel compliance
**D.** It is inversely related to the cross sectional are of a vessel

**Ans. 16. (B)**

Flow velocity is solely determined by blood flow and cross sectional area of the vessel, compliance change will affect the pulse wave velocity but not the flow velocity.

**17. Each of the following statements about the blood flow is true EXCEPT**

**A.** Is continuous in the arterioles that supply the microcirculation
**B.** Is continuous in the metarterioles of the micro-circulation
**C.** Is continuous in the capillary beds of the skeletal muscle
**D.** Is discontinuous in the true capillaries

**Ans. 17. (C)**

In resting skeletal muscle capillaries are in vasomotion.

**18. Find out the net filtration pressure if Cap. Hyd. Pr. is 20 mm Hg, Pl.Coll. Osm. Pr is 25 mm Hg, Int fl. Pr is 2 mm Hg, Coll. Osm. Pr. Of int. flu. is 5 mm Hg**

**A.** 2 mm Hg **B.** 5 mm Hg
**C.** –2 mm Hg **D.** –4 mm Hg

**Ans. 18. (C)**

Forces tending to filter the fluid = cap hyd pr + int coll osm press (20 + 5 =25), forces opposing the filtration = pl coll osm pr + int fl pr (25 + 2 =27). 25 – 27 = –2.

**19. Interstitial fluid osmotic pressure is**

**A.** Dependent on concentration of plasma proteins in interstitial space
**B.** Increased by lymphatic drainage
**C.** Proportional to the fluid present in the interstitial space
**D.** Opposes formation of edema

**Ans. 19. (A)**

Colloidal osmotic pressure depends on the concentration of plasma proteins in the fluid, lymph is one of the mechanism of returning the proteins in the interstitial space to blood.

**20. Formation of tissue fluid is increased when**

**A.** Capillary hydrostatic pressure decreases
**B.** Plasma protein decreases
**C.** Decreased venous pressure
**D.** There is a vasoconstriction

**Ans. 20. (B)**

**21. Average systemic capillary hydrostatic pressure is**

**A.** 2 mm Hg **B.** 60 mm Hg
**C.** 30 mm Hg **D.** 28 mm Hg

**Ans. 21. (C)**

**22. Vasomotor center is inhibited by**

**A.** Raised CSF pressure
**B.** Baroreceptor
**C.** Decreased blood pressure
**D.** Chemoreceptor

**Ans. 22. (B)**

Baroreceptor has inhibitory effect rest all stimulate the vasomotor center.

**23. Blood flow to the brain**

**A.** Changes with cardiac cycle
**B.** Varies with the change in blood pressure
**C.** Increases with increase in intracranial tension
**D.** Is autoregulated

**Ans. 23. (D)**

Blood supply to the brain is remarkably kept constant with well developed autoregulatory mechanism. It is kept constant with in the blood pressure of 60 mmHg to 160 mmHg.

**24. At rest brain receives**

**A.** 13-15% of cardiac output
**B.** 21% of cardiac output
**C.** 5% of cardiac output
**D.** 20% of cardiac output

**Ans. 24. (A)**

At rest brain receives 750 ml of blood.

**25. All of the following statement about pulmonary circulation are true EXCEPT**

**A.** Vessels are highly distensible
**B.** Has same resistance as the systemic arteries
**C.** Hypoxia increases the vascular resistance
**D.** Transmural pressure determines the caliber of the vessels

**Ans. 25. (B)**

Resistance of the pulmonary circulation is 1/10th of the systemic circulation, hypoxia causes the vasoconstriction of the pulmonary vessels.

**26. The cutaneous circulation**

**A.** Is primarily controlled by local metabolites
**B.** Increases in cold environment
**C.** Is primarily controlled by neural activity
**D.** Increases during anxiety

**Ans. 26. (C)**

Cutaneous vessel has sympathetic adrenergic vasoconstrictor innervation and sympathetic cholinergic vasodilator innervation (active during heat stress). During anxiety there is sympathetic stimulation, and blood is diverted to the other tissues (fright and flight reaction).

**27. All of the following statements about cutaneous circulation are true EXCEPT**

**A.** Acts as reservoir for blood
**B.** Helps in temperature regulation
**C.** Sympathetic vasoconstrictor fibres releases bradykinin
**D.** At rest receives 9% of the cardiac output

**Ans. 27. (C)**

Bradykinin is released by vasodilator system.

**28. All of the following statements about splanchnic circulation are true EXCEPT**

**A.** It receives 25% of cardiac output at rest

**B.** Blood flow increases following a heavy meal

**C.** Acts as an reservoir

**D.** Is controlled primarily by metabolites

**Ans. 28. (D)**

Splanchnic circulation is primarily under the control of sympathetic nerves.

**29. The normal pulmonary arterial pressure ranges from**

**A.** 120/80 mm Hg **B.** 27/10 mm Hg

**C.** 25/5 mm Hg **D.** 5/2 mm Hg

**Ans. 29. (B)**

**30. Primary stimulus to cerebral circulation is**

**A.** Hypoxia **B.** Hypercapnea

**C.** Increase pH **D.** Decreased pH

**Ans. 30. (D)**

Though $H^+$ ion cannot cross the blood-brain barrier in terms of sensitivity of vasomotor center, $H^+$ ion is the strongest stimulus.

# 3

# Nerve and General Physiology

**1. The term homeostasis means**

A. Arrest of bleeding
B. Maintenance of internal environment
C. Maintenance of blood pressure
D. Maintenance of blood flow

**Ans. 1. (B)**

Arrest of bleeding is called hemostasis, homeostasis is maintenance of internal equilibrium, i.e. ECF.

**2. The principle fluid medium of the cell is**

A. Water
B. Lymph
C. Blood
D. Plasma

**Ans. 2. (A)**

**3. Next to the water most the abundant substance in cells is**

A. Electrolytes
B. Golgi complex
C. Proteins
D. RBCs

**Ans. 3. (C)**

**4. A lipid bilayer of the cell is membrane is**

A. Fluid
B. Is miscible with the ICF
C. Solid
D. None of the above

**Ans. 4. (A)**

**5. Simple diffusion**

A. Occurs against electrochemical gradient
B. Occurs with the use of the energy
C. Occurs with the help of the carrier proteins
D. Is a passive process

**Ans. 5. (D)**

Simple diffusion occurs because of the kinetic motion of a particle and does not require any expenditure of energy.

**6. The rate of simple diffusion depends on all of the following EXCEPT**

**A.** Conc. difference of substance
**B.** No. of openings in the cell membranes
**C.** Velocity of the kinetic motion
**D.** Conc. of the carrier proteins in the cell membrane

**Ans. 6. (D)**

As simple diffusion is just the movement of particle which can easily cross the membrane it does not require any carrier protein.

**7. Which of the following would be expected to increase the permeability of the cell membrane**

**A.** Increased membrane thickness
**B.** Increased no. of the protein channels
**C.** Decreased lipid solubility
**D.** Decreased temperature

**Ans. 7. (B)**

Osmosis is the diffusion of water molecule from higher to lower concentration.

**8. Osmosis means movement of**

**A.** Water from high conc. to low conc.
**B.** Solute from high conc. to low conc.
**C.** Proteins from high conc. to low conc.
**D.** Water from low conc. to high conc.

**Ans. 8. (A)**

**9. The osmotic pressure of the solution**

**A.** Prevents the diffusion of water
**B.** Is proportional to the concentration of water
**C.** Is inversely proportional to the concentration of the solutes
**D.** None of the above

**Ans. 9. (A)**

Osmotic pressure is exerted by the particle which can not easily cross the membrane so it creats a pressure which holds back the water molecule.

**10. Osmolarity is**

**A.** Measured as osmoles/litre of solution
**B.** Measured as osmoles/kg of water

**C.** Same as osmolality
**D.** A better indicator of osmotic pressure

**Ans. 10. (A)**

When osmolar concentration is measured in kilogram it is called osmolality, osmolality and osmolarity have minor difference in the values (less than 1%).

**11. Primary active transport**

**A.** Utilizes a kinetic energy for movement of an ion or molecule
**B.** Uses ATP as a source of energy for movement of an ion or molecule
**C.** Does not have a saturation point
**D.** Moves an ion from higher to lower concentration

**Ans. 11. (B)**

Active process utilizes energy in the form of ATP to move the substance from lower to higher concentration, because it requires carrier it reaches a saturation point.

**12. A diffusion potential development in a membrane permeable to many ions depends on all of the following factors EXCEPT**

**A.** Polarity of the electric charges of each ion
**B.** Permeability of the membrane to each ion
**C.** Conc. difference of each ion across the membrane
**D.** No. of carrier proteins in cell membrane

**Ans. 12. (D)**

Diffusion potential depends on the normal diffusion coefficient and diffusibility of the molecule across membrane and hence carrier protein does not play a role.

**13. All of the following statements regarding rough endoplasmic reticulum are true EXCEPT**

**A.** It is composed of ribosome granules
**B.** It is primarily involved in protein synthesis
**C.** It is primarily involved in lipid synthesis
**D.** It has RNA

**Ans. 13. (C)**

Rough ER has ribosomes on its membrane which is contains RNA and enzymes and is involved in the synthesis of proteins. The smooth ER which lacks ribosomes is more involved in the synthesis of lipids.

**14. Which of the following does not determine the RMP**

A. $K^+$ diffusion potential

B. Na K pump

C. $Na^+$ diffusion potential

D. $Ca^{++}$ diffusion potential

**Ans. 14. (D)**

The permeability of the membrane for calcium is less compared to for sodium and potassium, most of the transport of calcium is through pump.

**15. Lysosomes are formed by**

A. Agranular ER

B. Granular ER

C. Golgi complex

D. Cell membrane

**Ans. 15. (C)**

Agranular and granular ER are concerned with lipid and protein synthesis.

**16. All of the following statement about mitochondria are true EXCEPT**

A. Enzymes are attached to the inner membrane

B. It is self replicative

C. It contains RNA

D. Are called the powerhouse of the cell

**Ans. 16. (C)**

Oxidative enzymes need for the formation of ATP are present on the inner membrane of the mitochondria, it has DNA which helps to replicate.

**17. The rate of pinocytosis depends on**

A. ECF calcium concentration

B. ECF sodium concentration

C. ICF potassium concentration

D. pH of the body

**Ans. 17. (A)**

The calcium reacts with the contractile protein filaments beneath the coated pits to provide the force for pinching the vesicles away from the cell membrane. The process of pinocytosis requires energy which is supplied by ATP.

**18. Peroxisomes are formed from**

A. Smooth endoplasmic reticulum

B. Nucleus

C. Rough endoplasmic reticulum

D. Cell membrane

**Ans. 18. (A)**

Peroxisomes are similar to lysosomes except that they are formed from smooth endoplasmic reticulum and contains oxidases rather than hydrolases.

**19. Which of the following transport processes uses ATP most directly**

**A.** Transport of sodium out of the cells
**B.** Transport of glucose into the cells
**C.** Transport of amino acid into the cells
**D.** Osmosis of water into the cell

**Ans. 19. (A)**

Glucose and amino acids are transported into cells by secondary active transport.

**20. Which of the following will reduce the resting membrane potential immediately**

**A.** Blocking the activity of sodium-potassium pump
**B.** Addition of isotonic KCL solution
**C.** Blocking $Cl^-$ channels
**D.** Replacing $Cl^-$ with $SO_4$

**Ans. 20. (B)**

Most of the contribution to RMP comes from potassium thus reducing the concentration difference will affect the RMP.

**21. A selective and irreversible increase in the sodium conductance without affecting potassium conductance would cause**

**A.** Hyperpolarization of the transmembrane potential
**B.** Transient hyperpolarization with return to normal potential
**C.** Transient depolarization with return to normal potential
**D.** Depolarization of transmembrane potential

**Ans. 21. (D)**

According to Goldman equation any increase in sodium conductance relative to potassium conductance will depolarize the membrane.

**22. The positive after potential in an AP**

**A.** Occurs at the end of depolarization
**B.** Is a misnomer

**C.** It is because of late opening of $Na^+$ channels
**D.** Is less negative than RMP

**Ans. 22. (B)**

It is a misnomer because the potential is actually more negative than the normal which is caused due to delayed closure of potassium channels. It is so called because earlier the measurement were done exteriorly.

**23. The velocity of conduction is faster in**
**A.** Myelinated nerve fibre
**B.** Small diameter nerve fibre
**C.** Cooled nerve fibre
**D.** Unmyelinated nerve fibre

**Ans. 23. (A)**

Velocity of conduction is proportional to the diameter of the nerve and to myelination. In myelinated nerves the action potential pumps from one node of the Ranvier to the other node of Ranvier because myelin sheath acts as an insulator.

**24. At the peak of the action potential**
**A.** The driving force for the sodium is greater than at the resting potential
**B.** The driving force for the potassium is less than at the resting potential
**C.** Sodium channels can not be activated
**D.** Conductance for the sodium is much less than that for potassium

**Ans. 24. (C)**

This is the absolute refractory period. conductance for sodium is still greater than potassium and transmembrane potential reverses sign at the peak of the action potential. The driving force for the potassium is highest and for sodium it is lowest.

**25. Which of the following would increase if the ECF concentration of sodium becomes one-half the normal**
**A.** Overshoot of the action potential
**B.** Sodium equilibrium potential ($E_{Na}$)
**C.** Rate of rise of the action potential
**D.** Resting potential

**Ans. 25. (D)**

Equilibrium potential for will reduce because of decrease in concentration difference. This will reduce the overshoot and also will slow the rate of rise of action potential because of decreased driving force on sodium and will cause a slight increase in membrane potential.

**26. The relative refractory period of nerve membrane is characterized by**

**A.** Partial recovery of sodium conductance
**B.** Complete inactivation of potassium conduction
**C.** A lower threshold for next action potential
**D.** A decreased potassium equilibrium potential

**Ans. 26. (A)**

Conductance for potassium and threshold for next action potential is higher.

**27. The myelin sheath**

**A.** Decreases the conduction velocity
**B.** Is interrupted by node of Ranvier
**C.** Increases the energy expenditure for membrane recovery
**D.** Decreases the relative refractory period

**Ans. 27. (B)**

Myelin sheath is the covering on the medullated nerve fibres, but it is not continuous. It is broken down at node of Ranvier. It acts as an insulator and action potential travels from node to node, thus increasing the conduction velocity. It helps to decrease the expenditure of energy for the recovery but has no effect on the refractory period.

**28. A local anesthetic agent**

**A.** Increases the sodium influx
**B.** Blocks sodium potassium pumps
**C.** Affects small diameter nerve fibres first
**D.** Promotes potassium efflux

**Ans. 28. (C)**

Local anaesthetic agent such as procaine blocks the voltage sensitive sodium channels. As the smaller fibres have less no. of sodium channels so greate fraction of channels are inactivated. But it has no effect on potassium conductance or efflux or on sodium potassium pump.

**29. The diffusion potential caused by potassium alone would give rise to a membrane potential of**

**A.** – 94 mv **B.** + 61 mv
**C.** – 90 mv **D.** – 4 mv

**Ans. 29. (A)**

According to Nernst equation diffusion potential for $K^+$ = –61 × log conc. of an ion inside/conc. outside, 35/1=35, log of 35 = 1.54, 1.54 × –61= –94

**30. The diffusion potential caused by sodium alone would give rise to a membrane potential of**

**A.** – 65 mv **B.** + 61 mv
**C.** + 90 mv **D.** – 4 mv

**Ans. 30. (B)**

Diffusion potential for $Na^+$ = – 61 × log conc. of an ion inside/conc. outside, 14/142 = 0.1, log of 0.1 × –61 = +61.

**31. The diffusion potential caused by sodium and potassium alone would give rise to a membrane potential of**

**A.** – 4 mv **B.** – 61 mv
**C.** + 61 mv **D.** – 86 mv

**Ans. 31. (D)**

The potential of the membrane permeable to more than one ion is calculated by using Goldman, Hodgkin, Katz, equation.

**32. Of the –90 mv resting membrane potential of the large nerve fibres, the sodium potassium pump contributes**

**A.** – 86 mv **B.** – 4 mv
**C.** – 60 mv **D.** – 10 mv

**Ans. 32. (B)**

Na. K pump is an electrogenic pump, which pumps 3 sodium out and 2 potassium in and hence causing loss of one positive ion which adds to the negativity of the membrane.

**33. Which of the following statement is TRUE**

**A.** RMP in the large skeletal muscle fibre is –60 mv

**B.** RMP in the small skeletal muscle fibre is –80 mv

**C.** RMP in the small nerve fibre is –100 mv

**D.** RMP of large nerve and skeletal muscle fibre is same

**Ans. 33. (D)**

The RMP of small skeletal and nerve fibres and some neurons of the CNS is –40 to –60 mv.

**34. All of the following statement about action potential are true EXCEPT**

**A.** $Na^+$ channel activation gate open at potential of –50 to –70

**B.** $Na^+$ channel inactivation gates are slow to close

**C.** $Na^+$ channel inactivation gate reopens at potential of 0 mv

**D.** $K^+$ channels are slow to open

**Ans. 34. (C)**

The same potential changes that opens the activation gate also closes Na channel inactivation gate opens potassium channel. But both of these are slow to react. The inactivation gates open at the original RMP or potential near to RMP.

**35. Decrease in ECF calcium**

**A.** Activates sodium channel

**B.** Inactivates sodium channel

**C.** Activates chloride channels

**D.** None of the above

**Ans. 35. (A)**

Decrease in ECF calcium has a profound effect on the voltage at which the sodium channel become activated. Fall in ECF calcium ion to 50% of the normal excites many peripheral nerves.

**36. The membrane fails to fire when the rise in the potential is gradual because**

**A.** Of failure of $Na^+$ channel activation gates to open

**B.** Of failure of $Na^+$ channel inactivation gates to close

**C.** $Na^+$ channel inactivation gates closes simultaneously with the opening of activation gates

**D.** Failure of opening of potassium channels at the same time as that of activation

**Ans. 36. (C)**

We know that inactivation gate of sodium channel is slow to open, but when the change in potential is gradual there is enough time for the inactivation gates to open, so the influx of the sodium is blocked.

**37. Action potential is characterised by all of the following EXCEPT**

**A.** It is nondecremental
**B.** Follows all or none phenomenon
**C.** It propagates in one direction only
**D.** It is self propagatory

**Ans. 37. (C)**

By definition action potential is a sudden change in the membrane potential which is self propagatory, nondecremental. It follows all or none phenomenon.

**38. When the CNS sends weak signal to contract a muscle the smaller motor neurons get excited first because**

**A.** Smaller motor neurons are more excitable than larger motor neurons
**B.** Smaller motor neurons are more in number
**C.** Smaller motor neurons have leaky $Na^+$ channels
**D.** Smaller motor neurons have leaky $K^+$ channels

**Ans. 38. (A)**

**39. Peripheral nerve has the ability to regenerate because of the presence of**

**A.** Neurilemma **B.** Axolemma
**C.** Perineurium **D.** Epineurium

**Ans. 39. (A)**

# 4

# Muscle

1. **The light band region of sarcomere**
   A. Is a region of intercalated disc
   B. Has only myosin filament
   C. Is a region of overlapping of actin and myosin filaments
   D. Has only actin filament

**Ans. 1. (D)**
Sarcomere is the structural and functional unit of muscle. It is composed of actin and myosin proteins. The intermingling of actin and myosin gives it a striated appearance. The light band is a region which is composed only of actin filament which is isotropic to the polarized light.

2. **Myosin molecule is composed of**
   A. 2 heavy chains and 4 light chains
   B. 2 light chains and 4 heavy chains
   C. 2 heavy chains and 2 light chains
   D. 4 heavy chains and 4 light chains

**Ans. 2. (A)**
Two heavy chains present in myosin winds spirally around each other forming a double helix. At one end the chain is folded to form a globular structure called head. Four light chains forms a part of the myosin head, two light chains for each head.

3. **Which of the following is not the part of the actin filament**
   A. Actin B. Troponin
   C. Tropomyosin D. Calmodulins

**Ans. 3. (D)**
Actin filament is composed of three protein components viz. actin, troponin, tropomyosin. Calmodulins is a calcium binding protein.

4. **Troponin subunits has a strong affinity for all of the following EXCEPT**

A. Actin
B. Calcium
C. Myosin
D. Tropomyosin

**Ans. 4. (C)**

Troponin protein is composed of three subunits, which has affinity for calcium, tropomyosin, and actin. Under resting state the acive sites present on actin molecule is covered by tropomyosin because of this affinity.

**5. The binding of the ATP to the myosin molecule causes**

A. Movement of myosin
B. Detachment of actin from myosin
C. Binding of actin and myosin
D. Uncovering of active sites

**Ans. 5. (B)**

Myosin head has a binding site for ATP and ATPase activity. This binding is responsible for the detachment of actin from myosin. The active sites are uncovered by binding of calcium to troponin.

**6. Maximum efficiency of muscle contraction is achieved**

A. When muscle contracts at a moderate velocity
B. When muscle contracts at a slower velocity
C. When muscle contracts at a faster velocity
D. Is independent of velocity of contraction

**Ans. 6. (A)**

Efficiency is the percentage of the energy that is converted into work instead of heat. When muscle contracts at a slow velocity (or without any movement) large amount of maintenance heat is released during contraction even though no work is performed, thus decreasing the efficiency. When contraction is too rapid large amount of energy is spent to overcome the viscous friction with in the muscle.

**7. Which of the following muscle protein has ATPase activity**

A. Tropomyosin
B. Myosin
C. Troponin
D. Actin

**Ans. 7. (B)**

Refer to Explanation of Ans 5.

**8. In the contraction of muscle, the word power stroke means**

**A.** Attachment of actin to myosin

**B.** Detachment of actin from myosin

**C.** Tilting of myosin head towards its arm

**D.** Attachment of ATP to myosin

**Ans. 8. (C)**

When the head of myosin binds to the active sites, this causes profound changes in the intramolecular forces between the head and arm of the cross bridges. The new alignment of the forces causes the head to tilt toward arm.

**9. The first source to reconstitute the ATP after ATP store has been depleted during muscle contraction is**

**A.** Glycogen

**B.** Oxidative metabolism

**C.** Phosphocreatine

**D.** Glucose

**Ans. 9. (C)**

**10. The concentration of ATP present in muscle fibre (about 4 m molar) is sufficient to maintain full contraction for**

**A.** 5–6 secs **B.** 8–10 secs

**C.** 1–2 secs **D.** 1 min

**Ans. 10. (B)**

**11. The ATP present in the muscle is used for all of the following EXCEPT**

**A.** Transport of sodium and potassium

**B.** Binding of actin to myosin

**C.** Detachment of myosin and actin

**D.** Pumping calcium in endoplasmic reticulum

**Ans. 11. (B)**

Transport of sodium and potassium by Na K pump helps to maintain RMP and thus excitability. Binding of ATP is necessary for the detachment of actin from myosin. After the contraction is over the calcium is transported actively back in the ER.

**12. Which of the following statement about fast muscle fibre is TRUE**

**A.** Has large no. of oxidative enzymes
**B.** Has poorly developed sarcoplasmic reticulum
**C.** Size is smaller
**D.** Has less extensive blood supply than slow muscle fibres

**Ans. 12. (D)**

Fast fibres contract rapidly and for a shorter period of time and requires fast supply of energy, which can be provided by glycolytic enzymes. Long-term blood supply is needed for sustained and longer contracting muscle.

**13. Which of the following is part of the series elastic component**

**A.** Hinged arms of cross bridges
**B.** Tendons
**C.** Sarcolemmal ends of muscle fibres where they attach to tendon
**D.** All of the above

**Ans. 13. (D)**

Series elastic components are those which do not contract during the muscle contraction against the load. The contractile part of the muscle must shorten extra 3–5% to make up for the stretch of these elements.

**14. Which of the following statement about slow muscle fibre is FALSE**

**A.** Has large no. of oxidative enzymes
**B.** Extensive blood supply
**C.** Fewer mitochondria
**D.** Presence of large amount of myoglobulin

**Ans. 14. (C)**

Slow fibres are the one which are involved in the long-term contraction, hence require continuous supply of blood and nutrients along with the oxidative enzyme and mitochondria.

**15. Before the exhaution of entire glycogen, it can supply the energy for the maximum period of**

A. 2–4 hours   **B.** 30 minutes
C. 5–7 hours   **D.** 10 minutes

**Ans. 15. (A)**

More than 95% of all the energy used by muscles for the sustained, long-term contraction is derived from oxidative metabolism. For the period of exercise lasting for 2–4 hours as much as one-half of the energy can come from stored glycogen before the glycogen store is depleted.

**16. The structural and functional unit of a muscle is**

**A.** Sarcolemma **B.** Sarcoplasm
**C.** Sarcomere **D.** Motor end plate

**Ans. 16. (C)**

Sarcollema is the membrane of the muscle fibre. Sarcoplasm is the cytoplasm of the muscle fibre. Motor end plate is the point where the nerve enters the muscle.

**17. The most excitable tissue in the body is**

**A.** Large myelinated nerve fibre
**B.** Skeletal muscles
**C.** Smooth muscles
**D.** Cardiac muscles

**Ans. 17. (A)**

**18. The seat of fatigue in intact body is**

**A.** Central synapse
**B.** Neuromuscular junction
**C.** Muscle itself
**D.** Nerve

**Ans. 18. (A)**

The seat of fatigue in nerve-muscle preparation is neuromuscular junction, but in the intact body to prevent the damage to the muscle the seat is in the synapses of CNS. The nerve itself is indefatiguable.

**19. An ACh receptor at the neuromuscular junction is a complex of**

**A.** 1α, 2 β, 1γ, 1δ **B.** 2α, 1 β, 1γ, 1δ
**C.** 1α, 1 β, 2γ, 1δ **D.** 1α, 1 β, 1γ, 2δ

**Ans. 19. (B)**

**20. Nicotine stimulates muscle fibre by**

**A.** Destroying acetylcholinesterase enzyme
**B.** Causing release of more Ach from postsynaptic terminal

C. Causing localized depolarization at motor end plate where ACh receptors are located

D. Causing influx of calcium into presynaptic terminal

**Ans. 20. (C)**

**21. The curariform drugs inhibits neuromuscular transmission by**

A. Inhibiting the release of ACh

B. Destroying ACh which is released

C. Increasing the concentration of Achesterase

D. Competitive inhibition of ACh

**Ans. 21. (D)**

Curariform drug competes with ACh for the ACh receptor located on the postsynaptic membrane (on the muscle membrane) the prevents the normal release of sodium from muscle membrane and thus its excitability.

**22. The drug neostigmine is helpful in myasthenia gravis because it**

A. Destroys the ACh receptor antibody

B. Destroys acetylcholinesterase

C. Causes release of more ACh

D. Causes increased release of calcium from SR

**Ans. 22. (B)**

In myasthenia gravis there are antibodies against acetylcholine receptor and hence the action of ACh is lessened. Neostigmine by destroying acetylcholinesterase increases the local concentration of the ACh in the synaptic cleft.

**23. In skeletal muscle fibre**

A. The duration of action potential is same as that of large myelinated nerve fibre

B. Action potential spreads to the interior through transverse tubules

C. RMP is – 60 mv to – 70 mv

D. Impulses travel at the speed of 30–40 m/sec

**Ans. 23. (B)**

Action potential in the skeletal muscle fibre lasts for 1–5 m/sec, five times longer than in large nerve fibres. The RMP is same as that of large nerve fibre about –80 to –90 mv. Conduction velocity is 1/18th of large nerve fibre about 3–5 m/sec.

**24. Maximal contraction of muscle occurs when there is**

**A.** Maximal overlapping of actin and cross bridges
**B.** Minimum overlapping of actin and cross bridges
**C.** Actin overlaps the opposite actin filament
**D.** None of the above

**Ans. 24. (A)**

The strength of contraction depends on the no. of cross bridges bound to active sites, which occurs when there is maximum overlap between actin and myosin filaments.

**25. Muscle contracts with the maximum strength when the sarcomere length is**

**A.** 1–1.5 micrometers
**B.** 2–2.2 micrometers
**C.** 2.5–3 micrometers
**D.** 3–3.5 micrometers

**Ans. 25. (B)**

Strength of muscle contraction is proportional to the no. of cross bridges bound to active sites. This depends upon the overlapping of actin and myosin which in turn depends on the resting length of the sarcomere. When the length is less than 2 micrometers actin of two opposite site starts overlapping each other. When the length is more than 2.2 micrometers there is no enough interaction between actin and myosin.

**26. All of the following statements about the muscle hypertrophies are true EXCEPT**

**A.** Increase in the number of muscle cells
**B.** Splitting of myofibrils
**C.** Increase in no. of actin and myosin filaments
**D.** It occurs when the muscle contracts against maximal force

**Ans. 26. (A)**

No. of muscle cell usually doesn't change after the birth, whatever hypertrophy occurs is because of incease in the size of the muscle fibre.

**27. After denervation, the muscle function can be restored if the nerve supply is restored within**

**A.** 5–6 months
**B.** 12–18 months

C. 2–3 months
D. 2 years

**Ans. 27. (C)**

Two months after the denervation, the degenerative changes begins to appear in muscle, if nerve supply is not restored within 3 months, than the chance of recovery of muscle function decreases progressively and by the end of 1-2 years total loss of ability to recover the muscle function.

**28. Rigor mortis**
A. Is caused due to inability of actin to bind myosin
B. Under normal condition lasts for 15 – 25 hours
C. Muscle rigidity is lost rapidly in cold climate
D. None of the above

**Ans. 28. (B)**

For detachment of the myosin head from the active site ATP has to bind to myosin head. Rigor mortis is sustained contracture following death due to unbinding of actin and myosin. After 15–25 hours lysosomes causes lysis of the muscle proteins. This lysis occurs rapidly at higher temperature.

**29. Which of the following statement about neuromuscular junction is TRUE**
A. The invaginating nerve fibre penetrates the muscle fibre plasma membrane
B. Motor end plate is covered by Schwann cell
C. Usually there are many neuromuscular junction per muscle fibre
D. Nerve invaginates muscle fibre at its periphery

**Ans. 29. (B)**

There is only one neuromuscular junction per muscle fibre which is at the center of the muscle fibre. The nerve lies outside the plasma membrane of the muscle fibre. The entire structure the invaginating nerve and point where it enters the muscle is called motor end plate which is covered by one or more Schwann cells that insulates it from the surrounding fluid.

**30. All of the following statement about acetylcholine receptor in the postsynaptic membrane is true EXCEPT**
A. It penetrates the entire membrane
B. It opens when acetylcholine binds to its alpha subunit

**C.** It allows sodium ion to pass through it

**D.** It allows chloride ion to pass through it

**Ans. 30. (D)**

Acetylcholine receptor is actually a acetylcholine gated ion channel which is made up of 2, subunit, one β, γ and δ subunits. When two ACh molecule binds to α subunit the channel opens and allow the passage of positive ions such as Na, K and Ca (more of Na). The interior of the channel being negative it doesn't allow negative ions to pass through it.

**31. The T tubule of the muscle**

**A.** Travel parallel to myofibrils

**B.** Are open to the exterior

**C.** Contains ICF in their lumen

**D.** Are present only at the surface of the membrane

**Ans. 31. (B)**

Transverse tubules originate from the cell membrane and extend all the way to the opposite side. It runs transverse to the myofibrils and is open to the exterior and hence contain ECF.

**32. The multiunit smooth muscle fibres**

**A.** Operates independently of each other

**B.** Is innervated by more than one nerve

**C.** Mainly controlled by non-neural stimuli

**D.** Exhibit spontaneous contractions

**Ans. 32. (A)**

The example of these type of muscles are ciliary muscle and iris of the eye, piloerector muscle.

**33. The unitary smooth muscle fibres**

**A.** Mean single muscle fibre

**B.** Is aggregated to form bundles

**C.** Does not have gap junction

**D.** Contract independently of each other

**Ans. 33. (B)**

Unitary muscle means a whole mass of many muscle fibres contract as a single unit. These fibres are aggregated to form bundles or sheets. Cell membrane have gap junction through which electrical activity in one muscle excites the entire bundle and all the muscles in that bundle contract together.

**34. The force of contraction from one cell to another in smooth muscle is transmitted through**
  **A.** Intercellular cytoplasmic continuation
  **B.** Bonds between the intercellular dense bodies
  **C.** Tight junctions
  **D.** None of the above

**Ans. 34. (B)**

Smooth muscle doesn't have striated appearance like that of skeletal muscle because of different pattern of distribution of actin and myosin. Actin is attached to the dense bodies. Some dense bodies are present in the membrane while some are dispersed inside the cell. There also is the bonding between some of the membrane dense bodies of the adjacent cell.

**35. Which of the following statement about smooth muscle contraction is NOT TRUE**
  **A.** The rate of cross bridge cycle is faster than in skeletal muscle
  **B.** Less energy is required to sustain the tension than skeletal muscle
  **C.** The force of contraction is greater than skeletal muscle
  **D.** Percentage of shortening is more than skeletal muscle

**Ans. 35. (A)**

The rate of attachment of actin to myosin and its subsequent detachment and again reattachment is slow. But the fraction of time that cross bridges remain attached to actin (force of contraction) is more in smooth musle. Smooth muscle can effectively contract to 2/3rd of its stretched length (skeletal m – ¼ –1/3rd).

**36. The latch phenomenon**
  **A.** Associated with increased expenditure of energy during contraction
  **B.** Allows long-term maintenance of tone in smooth muscle
  **C.** Allows long-term maintenance of tone in skeletal muscle
  **D.** Is because of rapid cross bridge cycling

**Ans. 36. (B)**

Latch phenomenon is present to a greater extent in smooth muscle. It allows the muscle to maintain contraction with the same amount of energy spent.

**37. The time required for relaxation of smooth muscle determined by**

**A.** Intracellular calcium concentration
**B.** Amount of active myosin kinase in the cell
**C.** Number of actin filaments in the cell
**D.** Amount of active myosin phosphatase in the cell

**Ans. 37. (D)**

The contractile process starts with increase in intracellular calcium level, which along with calmodulin activates enzyme myosin kinase. This enzyme in turn phosphorylates the regulatory chain light chain in myosin head. With decreased intracellular calcium all the processes are reversed except phosphorylation of head. The dephosphorylation is brought about by enzyme myosin phosphatase.

**38. Which of the following channel is important in generating action potential in smooth muscle**

**A.** Voltage gated $K^+$ channel
**B.** Voltage gated $Na^+$ channel
**C.** Voltage gated $Ca^{++}$ channel
**D.** Voltage gated $Cl^-$ channel

**Ans. 38. (C)**

Smooth muscle membrane have far more calcium channels than sodium channels, but this channels are slow to open which explains the slow action potential in smooth muscle.

**39. The calcium that initiates the contractile process in smooth muscle comes mainly from**

**A.** Sarcoplasmic reticulum
**B.** Extracellular fluid
**C.** Golgi complex
**D.** None of the above

**Ans. 39. (B)**

The sarcoplasmic reticulum of smooth muscle is poorly developed.

**40. The slow wave rhythm in the smooth muscle**

**A.** Can initiate an action potential
**B.** Are called pacemaker waves

**C.** Can not cause muscle contraction on its own
**D.** All of the above

**Ans. 40. (D)**

These waves are seen especially in intestinal smooth muscle. The exact cause for this is not known but is believed to be cause by rhythmic change in the conductance of the ion channels. It is a local potential.

# 5

# Body Fluids

**1. Water loss by lungs and skin in millilitres per day is**

A. 100 B. 700

C. 1400 D. 200

**Ans. 1. (B)**

This is called as insensible water loss. The loss through skin occurs independently of sweating and is present in person who is born without sweat glands. This loss through skin which is about 300–400 ml per day is minimized by cholesterol filled cornified layer. This loss becomes appreciable when this layer is denuded as in case of burns. The water loss through respiratory tract is about 300–400 ml per day and increases in cold weather.

**2. Water loss by faeces in millilitres per day is**

A. 100 B. 700

C. 1400 D. 200

**Ans. 2. (A)**

The daily water loss in faeces is very little, but increases in case of diarrhoea, hence people with severe diarrhoea have to be monitored for the fluid loss.

**3. Water loss by urine in millilitres per day is**

A. 100 B. 1400

C. 400 D. 700

**Ans. 3. (B)**

Water loss by kidney is usually about 1200–1400 ml per day. Kidney is the organ through which body tries to maintain the balance between input and output of fluid and electrolytes. Urine volume can vary from 0.5 lit/day in a dehydrated person to 20 lit/day in a person drinking excessive amount of fluid.

**4. Water loss by sweat in millilitres per day is**

A. 800 B. 100

C. 1200 D. 500

**Ans. 4. (B)**

The amount of fluid loss by sweat is highly variable. In the hot weather this can count for 1–2 lit/hour. This can deplete the body water rapidly if intake of fluid is not increases accordingly.

**5. In a 70kg adult human, the total body water averages about**

**A.** 40% **B.** 50%

**C.** 60% **D.** 70%

**Ans. 5. (C)**

In a normal 70 kg adult human being the total body water is about 42 litres that is around 60% of the total body weight. The percentage can change depending on the age, sex, and degree of obesity. Women contain slightly less water than men.

**6. The percentage of the total body weight contributed by water decreases with advancing age because of**

**A.** Increase in fat percentage of body

**B.** Decreased calcium in body

**C.** Decreased muscle mass

**D.** Decreased metabolism of the body

**Ans. 6. (A)**

As the age advances the percentage of the total weight contributed by fat increases. Decreased calcium is more related to osteoporosis.

**7. The ICF constitutes about**

**A.** 40% of the total body weight

**B.** 80% of the total body weight

**C.** 20% of the total body weight

**D.** 50% of the total body weight

**Ans. 7. (A)**

Out of the total 42 litres of fluid in the body about 28 litres is present inside the 75 trillion cells, which is about 40% of the total body weight. The fluid contains its individual mixture of different constituents, but concentration of these different substance is remarkably constant from one cell to another.

**8. The ECF constitutes about**

**A.** 40% of the total body weight

**B.** 50% of the total body weight

C. 20% of the total body weight
D. 80% of the total body weight

**Ans. 8. (C)**
All the fluid outside the cells are called ECF. It account for about 20% of the total body weight or 14 litres.

**9. Total body water can be measured by**
A. Antipyrine
B. Radioactive sodium
C. $^{125}$I-albumin
D. Evans blue

**Ans. 9. (A)**
Antipyrine is very lipid soluble and can easily penetrate the cell membranes and distribute uniformly throughout intracellular and extracellular compartments.

**10. Total volume of fluid in the extracellular compartment can be measured by**
A. Heavy water [$^2H_2O$, deuterium]
B. Radioactive water [$^3H_2O$, tritium]
C. Radioactive inulin
D. Evans blue dye [T – 1824]

**Ans. 10. (C)**
ECF volume can be measured by any substance which can disperse in plasma and interstitial fluid without penetrating the membranes. Usually some amount of the substance used to measure ECF volume penetrates the membrane, therefore we usually refer to this measurements as inulin space or sodium space etc.

**11. Which of the following will lead to intracellular edema**
A. Depression of the metabolic system of the tissue
B. Lack of adequate nutrition to the cells
C. Inflammation of the tissue
D. Increased permeability of capillaries

**Ans. 11. (D)**
Depressed metabolic system or inadequate nutrition causes the accumulation of sodium inside the cells, because of failure of pump. Inflammation increases the permeability of the membrane.

**12. All of the following can lead to accumulation of fluid in to extracellular space EXCEPT**

**A.** Decreased arteriolar resistance
**B.** Lymphatic blockage
**C.** Decreased blood flow to the tissue
**D.** Decreased plasma proteins

**Ans. 12. (C)**

Decreased blood flow to the tissue leads to decreased supply of the nutrients to the tissue which would result in intracellular edema.

**13. The safety factor for the prevention of accumulation of fluid in extracellular space is about**

**A.** 22 mm Hg **B.** 28 mm Hg
**C.** 32 mm Hg **D.** 17 mm Hg

**Ans. 13. (D)**

The total safety factor is about 17 mmHg which tries to prevent the formation of edema. Out of this 17 mmHg 3 mmHg is contributed by low tissue compliance, 7 mmHg is contributed by increased lymph flow and wash down of proteins from interstitial space accounts for remaining 7 mmHg.

**14. Increase lymphatic flow prevents the formation of edema by**

**A.** Increasing plasma colloid osmotic pressure
**B.** Decreasing capillary hydrostatic pressure
**C.** Decreasing interstitial colloidal osmotic pressure
**D.** None of the above

**Ans. 14. (C)**

Lymph is the only way through which proteins in the interstitial space is returned back to circulation. By washing away the proteins from interstitial space it decreases the net filtration of fluid from vessel to the interstitial space.

**15. The lymphatic channels are present in all of the following tissues EXCEPT**

**A.** Central nervous system **B.** Bones
**C.** Liver **D.** Spleen

**Ans. 15. (A)**

**16. Lymphatic channels are present in which of the following**

**A.** Superficial portions of the skin
**B.** Small intestines

C. Deeper portions of peripheral nerves
D. Endomysium of muscles

**Ans. 16. (B)**

Small intestines have lymphatic channels which is important for the transport of fats.

**17. All of the following statements regarding lymph flow are true EXCEPT**

A. Interstitial fluid from brain flows directly into blood vessels
B. Lymph from the lower part flows in to thoracic duct
C. Lymph from left side of head and arms empties into thoracic duct
D. Lymph from right side of head and arm empties into right lymphatic duct

**Ans. 17. (A)**

Brain is one of the few organs in the body which lacks lymphatic channels. But it has prelymphatic through which interstitial fluid from brain empties in to cerebrospinal fluid and thence directly back into the blood. lymphatics from head, upper body and lower body drains into thoracic duct and right lymphatic duct which ultimately empties into venous plexus.

**18. The total amount of lymphatic flow in a resting human averages about**

A. 20 millilitres per hour
B. 100 millilitres per hour
C. 80 millilitres per hour
D. 120 millilitres per hour

**Ans. 18. (D)**

About 100 millilitres of the lymph flows through the thoracic duct of a resting human per hour and about another 20 millilitres per hour flows through other lymphatic channels. Thus the total lymphatic flow is averages 120 millilitres per hour about 2–3 litres per day.

**19. The rate of lymphatic flow in the body is determined by**

A. Activity of the lymphatic pump
B. Interstitial fluid pressure
C. Both the above factors
D. None of the above

**Ans. 19. (C)**

The smooth muscles in the larger lymphatic vessels contracts automatically when stretched by fluid. Each lymphatic segment between valves contract independently of other. Some believe existence of the actinomyosin in endothelium of lymphatic capillary. Rise in the interstitial fluid pressure from negative value to about +1 to +2 increases the lymphatic flow, but pressure beyond this there is no further rise in the flow because of the external compression of the lymphatic vessels.

**20. What will be the volume of extracellular fluid if inulin administerd – 2.0 gm inulin excreted – 0.4 gm inulin concentration – 0.1 mg/ml**

| | |
|---|---|
| **A.** 16 litres | **B.** 20 litres |
| **C.** 1.6 litres | **D.** 0.6 litres |

**Ans. 20. (A)**

ECF volume is equal to the substance infused minus the substance excreted, divided by the concentration of the substance = 2.0 – 0.4 / 0.1 = 16 litres.

**21. The concentration of which of the following is different in plasma and interstitial fluid**

| | |
|---|---|
| **A.** Sodium ion | **B.** Small solutes |
| **C.** Protein | **D.** Chloride ion |

**Ans. 21. (C)**

All other can diffuse easily. But membrane is relatively impermeable to proteins.

# 6

# Excretory System

**1. The most important function of the kidney is**

A. Excretion
B. Homeostasis
C. Temperature regulation
D. Rennin secretion

**Ans. 1. (B)**

The most important function of kidney is homeostasis. Other functions are by the virtue of homeostatic function of kidney.

**2. Net filtration pressure in kidney (glomerulus) is**

A. 20 mm Hg
B. 25 mm Hg
C. 10 mm Hg
D. 30 mm Hg

**Ans. 2. (C)**

Net filtration pressure is the difference in the factor favoring the filtration and those opposing filtration.

**3. Normal renal blood flow per min is**

A. 1400ml
B. 1200ml
C. 1500ml
D. 1000ml

**Ans. 3. (B)**

Kidney is one of the organ which has high blood supply. It receives 19–21% of the cardiac output. Both the kidney together constitutes about 0.4% of the total body weight.

**4. All of the following statements regarding renal blood flow (RBF) are true EXCEPT**

A. RBF increases with increase in renal arterial pressure
B. RBF increases with increase in renal venous pressure
C. RBF decreases with increase in total renal vascular resistance
D. Renal arterial pressure is equal to the systemic arterial pressure

**Ans. 4. (B)**

Renal blood flow is determined by pressure gradient across the renal vasculature divided by the total renal vascular resistance. This equals to ( renal arterial pressure – renal venous pressure) / total renal vascular resistance. Renal arterial pressure is equal to the systemic arterial pressure and renal venous pressure averages 3 mmHg. The total vascular resistance is determined by the sum of the resistance in arteries, arterioles, capillaries and veins.

**5. Most of the resistance to the renal blood flow is offered by**

**A.** Efferent arterioles **B.** Afferent arterioles
**C.** Peritubular capillaries **D.** Renal vein

**Ans. 5. (A)**

Most of the renal vascular resides in interlobar arteries, afferent arterioles and the efferent arterioles. Under normal circumstances efferent arterioles, afferent arterioles, peritubular capillaries and renal vein are responsible for 43%, 26%, 10% and 0% of the total vascular resistance.

**6. Glomerular filtration is opposed by**

**A.** Colloidal osmotic pressure in peritubular capillaries
**B.** Colloidal osmotic pressure in glomerular capillaries
**C.** Hydrostatic pressure in efferent arterioles
**D.** Hydrostatic pressure in glomerular capillaries

**Ans. 6. (B)**

As everywhere else in the body the filtration is determined by the Starling's forces. Peritubular capillaries does not play a role in glomerular filtration. GFR is dependent on the forces acting in Bowman's capsule and glomerular capillaries.

**7. Renal autoregulation is influenced to a great extent by**

**A.** Epinephrine **B.** Acetylcholine
**C.** Angiotensin **D.** Nonc of the above

**Ans. 7. (D)**

Kidney has the intrinsic ability to regulate its blood flow without any external influences. It can effectively regulates its blood flow within the blood pressure of 80–170 mmHg.

**8. Glomerular filtration will decrease when there is**

**A.** Constriction of efferent arterioles
**B.** Dilatation of afferent arterioles

**C.** Obstruction in the urinary system
**D.** Decrease in plasma protein

**Ans. 8. (C)**

Constriction of efferent arteriole increases capillary hydrostatic pressure, dilatation of afferent arterioles increases the renal plasma flow. Decreased proteins will decrease colloidal pressure.

**9. Inulin is used to measure**
**A.** Renal blood flow
**B.** Renal plasma flow
**C.** Urinary volume
**D.** Glomerular filtration rate

**Ans. 9. (D)**

Inulin is a non-toxic polysaccharide that is neither reabsorbed or secreted by the tubules. Also it does not bind to plasma proteins and hence is easily filterd. Thus volume of plasma cleared of inulin per minute must equal the plasma filtered through glomerulus per minute.

**10. Which of the following substance is used to measure renal plasma flow**
**A.** Creatinine
**B.** Para-aminohippuric acid
**C.** Inulin
**D.** Radioactive iothalamate

**Ans. 10. (B)**

PAH is not only filtered but is also secreted into the renal tubules. Therefore PAH will be completely cleared from the plasma by renal excretion during a single circuit of plasma flow through the kidney. Thus all plasma supplying nephrons can be cleared of PAH. Normally 85–90% of the total plasma flowing through the kidney is cleared of PAH.

**11. Sodium can be transported across the luminal membrane of tubular cells by all of the following mechanism EXCEPT**
**A.** Counter transport with hydrogen ion
**B.** Co-transport with organic solutes
**C.** Sodium-potassium ATPase system
**D.** Through sodium channels

**Ans. 11. (C)**

Sodium-potassium ATPase pump is present at the baso-lateral surface of the renal tubules.

**12. Filtered chloride ions are reabsorbed by the process of secondary active transport in**

**A.** Ascending limb of loop of Henle
**B.** Proximal tubules
**C.** Distal tubules
**D.** Collecting ducts

**Ans. 12. (A)**

The only part of the renal tubule where sodium dependent secondary active transport of chloride ions occur is ascending of loop of Henle.

**13. The potassium in the renal tubule is**

**A.** Only filtered and secreted
**B.** Only filtered and reabsorbed
**C.** Filtrated, reabsorbed and secreted
**D.** Only secreted

**Ans. 13. (C)**

The potassium is freely filtered through the glomerulus and is then reabsorbed mainly by the proximal convoluted tubules. Then it is secreted by distal tubules and collecting ducts.

**14. In the loop of Henle**

**A.** Ascending limb reabsorbs little or no water
**B.** Descending limb reabsorbs active solutes
**C.** Ascending limb reabsorbs solutes as well as water
**D.** Ascending limb reabsorbs water

**Ans. 14. (A)**

The ascending limb of loop of Henle actively reabsorbs sodium and chloride but is relatively impermeable to water. The descending limb of loop of Henle is relatively impermeable to solutes and also does not actively reabsorb solutes.

**15. The primary active transport of $H^+$ takes place in**

**A.** Proximal convoluted tubules
**B.** Ascending limb of Loop of Henle
**C.** Early distal convoluted tubules
**D.** Late distal and cortical collecting tubules

**Ans. 15. (D)**

Late distal and cortical collecting tubules have intercalated cells which secretes $H^+$ by primary active transport mechanism. This mechanism is different from secondary active transport occurring in PCT. It can secrete $H^+$ ion against a large concentration gradient, as much as 1000 to 1. Thus these cells play important role in acid-base regulation.

**16. Loop of Henle's significantly contributes to the process of concentration and dilution of urine by**

**A.** Producing hyposmotic tubular fluid
**B.** Generating high osmotic gradient in the medullary interstitium
**C.** Generating high osmotic gradient for fluid within the loop
**D.** Participating in urea recycling

**Ans. 16. (B)**

Loop of Henle is the site for counter current multiplier. Though it generates high osmotic gradient with in the loop, participates in urea recycling and produces hyposmotic tubular fluid, for the formation of concentrated urine a significant contribution is by generation of high osmolar gradient in the medullary interstitium.

**17. Early part of DCT is referred to as diluting segment because**

**A.** It is impermeable to $Na^+$, $K^+$ and $Cl^-$
**B.** It is permeable to urea and water
**C.** Both are correct
**D.** None of the above

**Ans. 17. (D)**

The early part of the distal tubule is highly convoluted and has many almost the same absorptive characteristic as that of thick segment of ascending limb of loop of Henle. It reabsorbs most of the ions but is virtually impermeable to urea and water. This leads to the dilution of the tubular fluid.

**18. The final site for processing urine is (though it reabsorbs only 10% of filtered $Na^+$ and Water)**

**A.** Proximal convolutes tubule
**B.** Cortical collecting duct

C. Medullary collecting duct

D. Renal papilla

**Ans. 18. (C)**

Though medullary collecting duct reabsorbs only 10% of water and sodium, its ability to reabsorb water is controlled by ADH. Unlike the cortical collecting duct, medullary collecting duct is permeable to urea. Thus some of the tubular urea is reabsorbed into the medullary interstitium, helping to raise the osmolality in this region of the kidney and thus contributing to kidney's overall ability to form concentrated urine.

**19. Angiotensin II returns the blood pressure and ECF volume to normal by**

A. Secretion of aldosterone

B. Constricting afferent arterioles

C. By dilating efferent arterioles

D. Secretion of ADH

**Ans. 19. (A)**

Angiotensin II causes the vasoconstriction of most systemic vessels, bit doesn't affect the resistance of renal vessels. It causes the conservation of sodium and water in part by direct effect on kidney, but most pronounced effect on sodium and water reabsorption is through stimulating the release of aldosterone from adrenal cortex.

**20. Which of the following statement about medullary collecting duct is TRUE**

A. Its permeability to water is controlled by aldosterone

B. Its permeability to water is controlled by ADH

C. It is impermeable to urea

D. It is incapable of secreting $H^+$ against large concentration gradient

**Ans. 20. (B)**

Aldosterone has its action on the late distal and cortical collecting tubule, where it increases the reabsorption of sodium. Medullary collecting duct is permeable to urea and the intercalated cells present in it can secret $H^+$ ion against large concentration gradient.

**21. The substance used to measure GFR should**

A. Filter freely across glomerular membrane

B. Not be reabsorbed

C. Not be secreted
D. All of the above

Ans. 21. (D)

**22. ADH causes the reabsorption of water from**
A. Proximal convoluted tubule
B. Ascending loop of Henle
C. Descending loop of Henle
D. Distal tubule and collecting duct

Ans. 22. (D)

**23. Kidney can concentrate urine to a maximum osmolality of**
A. 120 mosmoles/lit
B. 500 mosmoles/lit
C. 1200 mosmoles/lit
D. 12000 mosmoles/lit

Ans. 23. (C)

The maximum osmolality of the medullary interstitium is 1200.

**24. Kidneys can regulate ECF $H^+$ ion concentration by**
A. Secretion of $H^+$
B. Reabsorbtion of filtered $HCO_3^-$ ion
C. Production of new $HCO_3^-$ ion
D. All of the above

Ans. 24. (D)

**25. The mechanism of $H^+$ ion secretion in early tubular segment is**
A. Secondary active transport
B. Facilitated diffusion
C. Primary active transport
D. Simple diffusion

Ans. 25. (A)

The $H^+$ ion is secreted in to the tubular lumen by secondary active transport which occurs in proximal convoluted tubules and primary active transport in late distal and cortical tubules and medullary collecting duct.

**26. Phosphate buffer plays important role in kidney because the**
A. pKa of phosphate buffer is near to the pH value of urine

**B.** As fluid becomes acidic more phosphate ions are formed

**C.** Phosphate ion helps in increasing the efficiency of the other buffers

**D.** None of the above

**Ans. 26. (A)**

The efficiency of the buffer is maximum when pH of the solution is near the pK of the buffer. The pK of the phosphate buffer is 6.8 and under normal circumstances pH of the urine is acidic, thus phosphate buffer is at its best efficiency range. Also phosphate gets concentrated in the tubule because of reabsorption of water and reabsorption of phosphate from the tubular fluid is poor.

**27. Ammonia buffer is important buffer in kidney because**

**A.** As the fluid becomes acidic the protein concentration in the kidney rises

**B.** The acidic pH stimulates enzyme glutaminase

**C.** The concentration of ammonia in kidney is normally high

**D.** Ammonia increases bicarbonate concentration in tubules

**Ans. 27. (B)**

The chronic acidosis stimulates enzyme glutaminase which causes increased formation of ammonium ($NH_4^+$ ion) from amino acid glutamine.

**28. The first micturition reflex is initiated when the volume of urine in urinary bladder is**

**A.** 150 ml **B.** 800 ml

**C.** 400 ml **D.** 50 ml

**Ans. 28. (A)**

The first micturition reflex is initiated when the amount of urine in the urinary bladder is 150 ml. Subsequently the rate of micturition waves increases and when the volume of urine is 400 ml, a strong desire to micturition arises. If the environment is not conducive the emptying of the bladder is voluntarily stopped. But when the volume of urine rises to 800 ml, the micturition reflex becomes painful and voluntary inhibition of emptying is no longer possible and the bladder voids the urine.

**29. The site of action of osmotic diuretics is**

A. Proximal convoluted tubule
B. Thick ascending loop of Henle
C. Early distal tubule
D. Collecting tubules

**Ans. 29. (A)**

**30. The site of action of loop diuretics is**

A. Proximal convoluted tubule
B. Thick ascending loop of Henle
C. Early distal tubule
D. Collecting tubules

**Ans. 30. (B)**

**31. The site of action of thiazide diuretics is**

A. Thick ascending loop of Henle
B. Collecting tubules
C. Proximal convoluted tubule
D. Early distal tubule

**Ans. 31. (D)**

**32. The site of action of aldosterone inhibitors is**

A. Proximal convoluted tubule
B. Thick ascending loop of Henle
C. Collecting tubules
D. Early distal tubule

**Ans. 32. (C)**

**33. The bicarbonate-carbon dioxide buffer is the most important physiological buffer in the body because**

A. Its pK is near to the normal pH of the body fluids
B. Level of $HCO_3$ and $CO_2$ in the body can be regulated
C. It is the only buffer present in the body
D. Its concentration in the ICF if high

**Ans. 33. (B)**

The pK of $HCO_3$ - $CO_2$ buffer is 6.1 and the normal pH of the body is between 7.35 – 7.4. The level of $HCO_3$ can be regulated by kidney while that of the $CO_2$ can be regulated by respiratory system. There are other buffers such as phosphate buffer, ammonia buffer, protein buffer present in the body besides bicarbonate buffer.

**34. An increased pH of the body will result in**

**A.** Excretion of acidic urine

**B.** Increase reabsorption of bicarbonate by kidneys

**C.** Decreased rate of ventilation

**D.** Increased activity of enzyme glutamase

**Ans. 34. (C)**

Increased pH is the state of alkalosis. In alkalosis the body tries to conserve $H^+$ ion and excrete bicarbonate ions making urine alkaline. The normal stimulation to the respiratory center is increased $CO_2$ level in body. Glutamase is stimulated by acidic pH.

**35. The pH of the body is regulated by the kidney by all of the following mechanism EXCEPT**

**A.** Secreting the hydrogen by primary active mechanism

**B.** Reabsorption of the filtered bicarbonate ions

**C.** Reabsorption of hydrogen ion by primary active transport

**D.** Excretion of an ammonium salt

**Ans. 35. (C)**

Hydrogen ion is actively secreted by intercalated cells in the collecting ducts. In the early tubular segment $H^+$ ion is reabsorbed by secondary active mechanism. Bicarbonate ions are reabsorbed in the PCT.

**36. The excretion of hydrogen ion in the urine will increase in**

**A.** Increased synthesis of ammonia by kidney

**B.** Increased reabsorption of $HCO_3$ by proximal convoluted tubules

**C.** Decreased level of phosphate buffers in kidney

**D.** Decreased pressure of carbon dioxide in blood

**Ans. 36. (A)**

Increased buffering capacity of the tubular fluid will lead to increase in hydrogen ion concentration. Increased reabsorption of $HCO_3$ ions in PCT doesn't increase $H^+$ ion excretion.

**37. The secretion of hydrogen ion by the proximal convoluted tubules will decrease in**

**A.** Increased filtration of $HCO_3^-$

**B.** Inhibition of carbonic anhydrase

**C.** Increased partial pressure of $CO_2$ in blood

**D.** Increasing filtered sodium

**Ans. 37. (B)**

Carbonic anhydrase is required for the formation of $H_2CO_3$. This carbonic acid then disassociate in to $H^+$ ion and $HCO_3^-$ ion. $CO_2$ is also required for the synthesis of $H_2CO_3$. ($CO_2 + H_2O$ *carbonic anhydrase* $H_2CO_3$.

**38. A patient with chronic lung disease is likely to suffer from**

**A.** Metabolic alkalosis
**B.** Respiratory acidosis
**C.** Respiratory alkalosis
**D.** Metabolic acidosis

**Ans. 38. (B)**

Chronic lung disease affects the transport of gases. This leads to accumulation of carbon dioxide in body, which leads to decrease in the pH of the body.

**39. Patient suffering from respiratory acidosis would have**

**A.** Decreased level of hydrogen ions in arterial blood
**B.** Increased level of dissolved $CO_2$ in arterial blood
**C.** Decreased $HCO_3^-$ level in arterial blood
**D.** Increased level of $HCO_3$ in arterial blood

**Ans. 39. (D)**

Increased $CO_2$ in the blood would increase the level of $HCO_3$ ions to counter the acidotic state.

**40. A person suffering from chronic respiratory disease with diarrhea will develop**

**A.** Respiratory acidosis with metabolic acidosis
**B.** Respiratory acidosis with metabolic alkalosis
**C.** Respiratory alkalosis with metabolic acidosis
**D.** Respiratory alkalosis with metabolic alkalosis

**Ans. 40. (A)**

Respiratory problems leads to accumulation of $CO_2$ in the body resulting in respiratory acidosis. Diarrhea leads to excretion of $HCO_3$, resulting in metabolic acidosis.

# 7

# Respiration

1. **Normal quiet expiration is brought about by**
   **A.** External intercostals muscle
   **B.** Internal intercostal muscle
   **C.** Elastic recoil of lung
   **D.** Abdominal muscle

**Ans. 1. (C)**

The normal expiration is a passive process and does not require support of any muscle to expire out the tidal volume. It is brought by the elastic recoil of lung tissues.

2. **Most of the work during inspiration is done by**
   **A.** Diaphragm
   **B.** External intercostals muscle
   **C.** Internal intercostal muscle
   **D.** Sternocleidomastoid muscle

**Ans. 2. (A)**

Diaphragm is responsible for the 60% of the inspiratory work. Remaining is brought about by external intercostal muscle. Sternocleidomastoid is an accessory muscle for inspiration. Internal intercostal is the muscle which helps during forceful expiration.

3. **The normal intrapleural pressure at the beginning of the inspiration is**
   **A.** 0 cm of $H_2O$
   **B.** –4 to –5 cm of $H_2O$
   **C.** –8 cm of $H_2O$
   **D.** +4 cm of $H_2O$

**Ans. 3. (B)**

Intrapleural pressure is the pressure of the fluid between the two layers of pleura (visceral pleura and parietal pleura). With the expansion of the chest during inspiration the pleural pressure changes from normal –4 to –5 cm of water to –7 to –8 cm of water. This –4 to –5 cm pressure is the pressure that is required to hold the lung open during resting state.

**4. Surfactant is a chemical substance which helps to**

**A.** Lower the surface tension
**B.** Bring about the closure of the alveoli
**C.** Relax the bronchial wall
**D.** Increase the work of breathing

**Ans. 4. (A)**

Surfactant is the surface active agent which reduces the surface tension at air water interface. In the lung it lines the alveoli and prevents the collapse of the lungs and helps to keep the lung open and thus reduces the work of breathing.

**5. Which of the following is not the component of the surfactant**

**A.** Phospholipid
**B.** Calcium
**C.** Apoprotein
**D.** Magnesium

**Ans. 5. (D)**

Surfactant is composed of phospholipid (dipalmityl phosphatidylcholine), apoprotein and calcium. Phospholipid helps to reduce the surface tension because of its hydrophobic (facing air) and hydrophilic (facing water) ends. Calcium helps in uniform spreading of surfactant.

**6. The premature babies have high collapsing alveolar pressure because**

**A.** The concentration of surfactant is less
**B.** The surfactant is immature
**C.** Alveoli are bigger in size
**D.** None of the above

**Ans. 6. (B)**

Surfactant is synthesized by type II alveolar epithelial cells. This cell constitutes 10% of the total cells. The recent research has shown that the concentration of surfactant in premature baby is normal, it is there maturity level that is less. After 7th month the maturation is brought about by different factors.

**7. All of the following help in lending stability to the alveoli EXCEPT**

**A.** Positive intrapleural pressure
**B.** Surfactant

C. Fibrous tissue of parenchyma

D. Interdependence of adjacent alveoli

**Ans. 7. (A)**

Surfactant gets concentrated in smaller alveoli and thus prevent further reduction in its size. The fibrous tissue and interdependence exert traction force on the adjacent alveoli. Thus helps to maintain the stability of the adjacent alveolar wall. Positive pleural pressure opposes the expansion of lung.

**8. Compliance of the lung and thorax together is**

A. More than the compliance of the lungs alone

B. Less than the compliance of the lungs alone

C. Same as that of the lungs alone

D. More than the compliance of the thorax alone

**Ans. 8. (B)**

The compliance of lung-thorax together is 110 ml/cm of water, while that of the lung alone is 200 ml/cm of water. This is because lungs and thorax to certain extent oppose the expansion of each other.

**9. To initiate the inspiration the work has to be done to overcome**

A. Compliance of lung and chest

B. Tissue resistance

C. Airway resistance

D. All of the above

**Ans. 9. (D)**

In order to expand the lungs the work has to be done to overcome compliance (surface tension, elastic tissue resistance), viscosity of lung and the chest wall structures (tissue resistance work) and airway resistance. Under normal quiet breathing most of the energy is spent to expand the lungs (compliance), a little energy is spent to overcome the viscosity of the lungs and a little more energy than tissue resistance is spent to overcome airway resistance.

**10. Which of the following increases during strenuous breathing**

A. Compliance  B. Airway resistance

C. Tissue resistance  D. Surface tension

**Ans. 10. (B)**

During strenuous exercise the velocity of the air flow through respiratory passage increases tremendously. This leads to the turbulence of air flow, which increases the resistance to air flow.

**11. The percentage of the energy spend by the body to bring about normal quiet ventilation is**

**A.** 40 - 50% **B.** 20 - 30%
**C.** 3 - 5% **D.** 10 - 20%

**Ans. 11. (C)**

During heavy exercise this can increase by 50 fold.

**12. Tidal volume under a normal physiological condition is**

**A.** 5800 **B.** 500
**B.** 4600 **D.** 1100

**Ans. 12. (B)**

Tidal volume is the amount of air inspired or expired out during normal quiet breathing. It is 500 millilitres.

**13. The amount of air in millilitres that can be forcefully expired out over and above normal tidal expiration is**

**A.** 3000 **B.** 4500
**C.** 800 **D.** 1100

**Ans. 13. (D)**

Amount of air that can be expired out over and above the normal tidal expiration is called expiratory reserve volume. This amounts to 1100 ml.

**14. The amount of air that remain in the lung after a forceful expiration is called**

**A.** Residual volume
**B.** Expiratory reserve volume
**C.** Functional residual capacity
**D.** Dead space

**Ans. 14. (A)**

Residual volume is that air which can not be expired out even after forceful expiration. It is this air that participates in gas exchange. Its normal value is 1200 ml.

**15. The maximum amount of air expired after a maximum inspiration is called**

**A.** Total lung capacity **B.** Vital capacity
**C.** Timed vital capacity **D.** Compliance

**Ans. 15. (B)**

Vital capacity is the maximum air that can be expired out following a maximum inspiration. It equals IRV + TV + ERV= 2800 + 500 + 1100 = 4400 ml. It is more in males than in females. It is more in standing and reclining posture than in lying down posture.

**16. The amount of air forcefully expired out per unit time after a forceful inspiration is called**

**A.** Total lung capacity **B.** Vital capacity
**C.** Timed vital capacity **D.** Compliance

**Ans. 16. (C)**

When the air is expired out forcefully, following a maximal forceful inspiration and is calibrated in terms of time, it is called timed vital capacity. This helps to differentiate obstructive and restrictive diseases of the lung.

**17. Which of the following can NOT be measured by Hutchinson's spirometer**

**A.** Inspiratory reserve volume
**B.** Expiratory reserve volume
**C.** Tidal volume
**D.** Residual volume

**Ans. 17. (D)**

**18. Alveolar ventilation means the amount of air flowing through**

**A.** Bronchus
**B.** Only alveoli
**C.** Gas exchange area
**D.** Entire respiratory tract

**Ans. 18. (C)**

Alveolar ventilation is the amount of air flowing through the gas exchange areas. The gas exchange area includes alveoli, alveolar sacs, alveolar ducts and the respiratory bronchioles. However during normal quiet breathing the volume of air in the tidal air is only enough to fill the respiratory passages down as far as the terminal bronchioles, with only a small portion of the inspired air actually flowing all the way in to the alveoli.

**19. Surfactant is secreted by**

**A.** Endothelium of the pulmonary capillary
**B.** Type II alveolar epithelial cells
**C.** Bronchial cells
**D.** Type I alveolar epithelial cells

**Ans. 19. (B)**

Capillary endothelium contains angiotensin converting enzyme.

**20. In a normal healthy person**

**A.** Anatomical and physiological dead space are nearly equal

**B.** Anatomical dead space is more than physiological dead space

**C.** Physiological dead space is more than anatomical dead space

**D.** None of the above

**Ans. 20. (A)**

Dead space is that part of the respiratory passage which does not participate in gas exchange, e.g. nose, pharynx, trachea. Besides this presence some non-functional alveoli also constitutes dead space. The earlier one is called as Anatomical dead space, which normally measures to about 150 ml. The latter constitutes alveolar dead space. Normally all the alveoli are functional. Thus alveolar dead space is zero. Anatomical and alveolar dead space together constitutes Physiological dead space. Since normally alveolar dead space is zero, the anatomical and physiological dead space are equal in value.

**21. Under normal respiratory conditions the greatest resistance to air flow occurs in**

**A.** Alveolar duct **B.** Nose

**C.** Large bronchi **D.** Terminal bronchi

**Ans. 21. (C)**

There are very few large bronchi and the hence resistance offered by them under normal physiological condition with normal respiratory functioning is more than the resistance offered by terminal bronchioles which number about 6500. Through each of these bronchi only a minute amount of air has to pass.

**22. When the volume of the air in the lung is at functional residual capacity level**

**A.** Lungs does not exert recoil pressure

**B.** Lungs exert recoil pressure toward lower volume

**C.** Pleural pressure is more than alveolar pressure
**D.** Chest wall exert recoil pressure toward lower volume

**Ans. 22. (B)**
At the FRC level lungs recoil toward the lower volume While chest recoil towards higher volume. At this volume of air in the lungs the pleural pressure (negative) is less than alveolar pressure (0).

**23. During inspiration**
**A.** Intrapleural pressure measures the recoil tendency of lungs, tissue resistance, airway resistance
**B.** Alveolar pressure should be more than atmospheric pressure
**C.** Lung volumes must always be above functional residual capacity
**D.** Air flow results due to elastic properties of lungs

**Ans. 23. (A)**
The alveolar pressure has to be less than the atmospheric pressure for the air to flow in. Inspiration can occur at any volume. Inspiration is an active process brought by the effort of diaphragm and external intercostals muscles.

**24. Most of the carbon dioxide in the blood is transported as**
**A.** Dissolved form
**B.** Bound to chloride
**C.** Bicarbonate ion
**D.** Carbamino compound

**Ans. 24. (C)**
Most of the $CO_2$ (70% ) in the body is transported in the form of $HCO_3^-$. When enzyme carbonic anhydrase is inhibited the transport of $CO_2$ is affected. In dissolved form only 7% is transported, the rest is transported in combination with hemoglobin and plasma proteins.

**25. Starting from trachea to alveoli there are**
**A.** 10-16 generations **B.** 20-25 generations
**C.** 26-30 generations **D.** 5-10 generations

**Ans. 25. (B)**
The air is distributed to lungs by the way of trachea, bronchi and bronchioles. Trachea is called the first

generation respiratory passage, two large bronchi are called the second generation respiratory passage. Each division after that is an additional division. There are about 20 to 25 generations before the air reaches alveoli.

**26. During quiet expiration**

**A.** Airway compression can occur if intrapleural pressure is more than airway pressure
**B.** Airway resistance is higher at greater lung volumes
**C.** Muscular effort is required
**D.** Air compression is more likely to occur at high lung volumes

**Ans. 26. (A)**

Airway compression is less likely to occur at higher lung volumes because a) elastic recoil of the lung is maximum b) airway resistance is minimal. Normal quiet expiration is brought about by elastic recoil of the lung.

**27. Arrange the following events in an order that they occur in cough reflex, i. closure of epiglottis, ii. entry of air, iii. sudden opening of epiglottis and vocal cords, iv. forceful contraction of abdominal muscles**

**A.** i, ii, iii, iv
**B.** ii, i, iii, iv
**C.** iv, i, iii, ii
**D.** i, iii, iv, ii

**Ans. 27. (B)**

**28. The sneeze reflex is caused by irritation in**

**A.** Lower respiratory passage
**B.** Alveoli
**C.** Terminal bronchiole
**D.** Nasal passages

**Ans. 28. (D)**

The sneeze reflex is caused by the irritation in the nasal passages. The afferent is carried by the 5th nerve to the medulla.

**29. Which of the following does not participate in articulation**

**A.** Larynx
**B.** Lip
**C.** Mouth
**D.** Tongue

**Ans. 29. (A)**

Larynx is especially adapted to act as vibrator. It helps in the act of phonation.

**30. Which of the following is NOT a function of the nose**

**A.** Warming of air

**B.** Humidification of air

**C.** Phonation

**D.** Filtration of air

**Ans. 30. (C)**

The air is warmed by the extensive surface of the conchae and septum. The warming , humidification and filtering of air together is called as airconditioning function of the upper respiratory passages. Ordinarily the inspired air rises to within 1°F of the body temperature and within 2 to 3% of full saturation with water before it reaches trachea. Breathing through trachea can lead to serious lung crusting and infection because of cooling and the drying effect especially in the lower lung.

**31. Lung acts as a reservoir for blood and has about**

**A.** 9-11% of total circulatory blood volume

**B.** 18-20% of total circulatory blood volume

**C.** 2-3% of total circulatory blood volume

**D.** 30% of total circulatory blood volume

**Ans. 31. (A)**

The quantity of the blood in lungs can vary from as little as one-half upto twice the normal. The total volume of blood in lungs is about 450 ml (9%) of the total blood volume. About 70 ml of this is in capillaries and the rest is equally divided between the arteries and the veins.

**32. In response to the decrease $PO_2$ conc. (< 70mm Hg) in alveoli the adjacent blood vessels**

**A.** Starts dilating

**B.** Starts constricting

**C.** Is not affected

**D.** Degenerates

**Ans. 32. (B)**

When the concentration of oxygen in alveoli decreases below normal, especially below 70 mmHg, the adjacent blood vessels starts constricting (opposite to the effect in systemic blood vessels), during the next 3 to 10 minutes. This is an important effect which allows the shunting of the blood to the well aerated alveoli.

33. **In normal upright adult, the lowest point in the lungs is**
    A. 40 cm below the highest point
    B. 60 cm below the highest point
    C. 30 cm below the highest point
    D. 100 cm below the highest point

**Ans. 33. (C)**

This is because of the effect of the gravitational force. This represents a 23 mmHg pressure difference, about 15 mm Hg of which are above the heart level and 8 mm Hg below the heart. Thus the pulmonary arterial pressure in the uppermost part of the lung in a standing person is 15 mm Hg less than the pulmonary arterial pressure at the heart level, and the pressure in the lowest portion of the lung is 8 mmHg higher than at the heart level.

34. **Regarding perfusion of lung, the Zone 1**
    A. Does not receive blood during any part of cardiac cycle
    B. Receives blood only during systole of heart
    C. Receives blood during entire duration of cardiac cycle
    D. Receives blood only during diastole of heart

**Ans. 34. (A)**

The capillaries in the alveolar walls are distended by the blood pressure within them, but at the same time they are also compressed by the alveolar pressure from outside. That is why whenever the alveolar pressure becomes greater than the pressure in the capillaries the capillaries close down and the blood flow through the capillaries stops. Zone 1 if present (as in positive pressure breathing in upright posture or hypovolumic states will be at the top most portion of the lungs, where the alveolar pressure is more than the capillary pressure. Normally in the lungs only Zone 2 and 3 exists.

35. **All of the following are part of the respiratory membrane EXCEPT**
    A. Layer of fluid and surfactant
    B. Layer of alveolar epithelium
    C. Fenestra
    D. Capillary endothelium

**Ans. 35. (C)**

The fenestra are the gaps in the capillary endothelium. These are present in the glomerular capillaries. The overall thickness of the respiratory membrane averages 0.6 micrometer. The total area of the respiratory membrane in an normal adult is 70 sq. meters.

**36. The oxygen carrying capacity at rest in a normal healthy individual is**

**A.** 29-31 ml/100ml of blood
**B.** 19-21 ml/100ml of blood
**C.** 19-21 lit/100ml of blood
**D.** 9-11 ml/100ml of blood

**Ans. 36. (B)**

The oxygen in the body is mostly transported by hemoglobin. Each gram of hemoglobin can carry 1.34 ml of oxygen. The normal level of hemoglobin is 14.5 – 16.5 gm%.

**37. The $PO_2$ of the arterial blood is**

**A.** 104 mm Hg **B.** 159 mm Hg
**C.** 40 mm Hg **D.** 95 mm Hg

**Ans. 37. (D)**

$PO_2$ of the atmospheric air is 159 mm Hg. But when this air reaches it is humidified that is it is diluted by the addition of water vapor pressure ($PO_2$ becomes 149 mmHg). At the normal tidal breathing the about 2300 ml of air is present in the lung at the end of normal expiration. The inspired air (350 ml) with the $PO_2$ of 149 mm Hg gets mixed with the 2300 ml of air (with $PO_2$ of 40 mm of Hg in the alveoli. This changes the $PO_2$ to 104 mmHg) at this pressure the oxygen enters the pulmonary circulation but later with the admixture from bronchial blood (deoxygenated) the $PO_2$ changes to 95 mmHg.

**38. Pulmonary circulation**

**A.** Is richly innervated by parasympathetic NS
**B.** Receives entire cardiac output
**C.** More at the top of the lungs in upright posture
**D.** High pressure circulation

**Ans. 38. (B)**

Pulmonary circulation is innervated by sympathetic as well as parasympathetic role. But sympathetic plays a bigger role than parasympathetic nerves. It is more at the bottom because of the effect of gravity. It is a low pressure circulation with arterial pressure of about 25/8 mmHg. It receives the entire blood from the right ventricle.

**39. Dorsal respiratory group of neurons are responsible for**

**A.** Initiation of inspiration
**B.** Initiation of expiration
**C.** Hering Breuer reflex
**D.** Apnoea

**Ans. 39. (A)**

Dorsal respiratory group of neurons extends along the entire length of medulla. These neurons emit repetitive bursts of inspiratory action potentials. These neurons decides the basic inspiratory pattern and rhythm. Hering-Breuer reflex is initiated by stretch receptors in the wall of the lungs.

**40. Which of the following statement regarding Hering-Breuer inflation reflex is NOT TRUE**

**A.** It is a protective reflex
**B.** Is initiated by stretch receptors of lungs
**C.** Is present at the normal tidal breathing
**D.** It arrests the inspiration

**Ans. 40. (C)**

Hering-Breuer reflex is a protective reflex initiated by the stretch receptors in the walls of the lungs. It prevents the extra inflation of the lungs and comes into the picture only when the tidal volume exceeds 1500 ml.

**41. Oxygen dissociation curve would be shifted to right when there is**

**A.** Decrease in the pH of the arterial blood
**B.** Decreased $PCO_2$ of the arterial blood
**C.** Decreased level of 2,3-DPG in blood
**D.** Excess of foetal haemoglobin

**Ans. 41. (A)**

All other factors shifts the oxygen dissociation curve to the left. Shift to right means the oxygen can be delivered to the tissue at the higher oxygen pressure.

**42. Oxygen dissociation curve would be shifted to left in all of the following conditions EXCEPT**

**A.** Presence of foetal hemoglobin
**B.** Decreased temperature

**C.** Increase pH
**D.** Exercise

**Ans. 42. (D)**

Exercise shifts the $O_2$ dissociation curve to right, because of decreased pH, increased temperature.

**43. Chronic cigarette smoking usually results in all of the following EXCEPT**

**A.** Increased level of carbon monoxyhaemoglobin in blood
**B.** Decreased pulmonary elasticity
**C.** Increased anatomical dead space
**D.** Emphysema

**Ans. 43. (C)**

Smoking acts as an the irritant to the bronchial mucosa which leads to bronchoconstriction. Bronchoconstriction decreases the dead space. Smoking decreases the activity of the enzyme $\alpha_1$ antitrypsin. This results in breakdown of the elastic tissue & alveolar septa causing emphysema.

**44. Work done to overcome airway resistance increases in**

**A.** Pulmonary tuberculosis
**B.** Vocal cord paralysis
**C.** Bronchial asthma
**D.** Saline inflated lungs

**Ans. 44. (C)**

Airway resistance increases in obstructive type of diseases. TB is a restrictive type of disorder. Saline inflated lungs results in decreased surface tension because of no air fluid interface. In Bronchial asthma there is bronchospasm which increases the resistance to air flow during expiration.

**45. Pulmonary arteries constriction results from activation of**

**A.** $\alpha_2$ adrenergic receptors
**B.** $ET_A$ endothelin receptors
**C.** $M_3$ muscarinic receptors
**D.** $H_2$ histaminic receptors

**Ans. 45. (B)**

$ET_A$ endothelin receptors are specific for endothelin I. This endothelin I is present in many tissues and mediates vasoconstriction.

**46. Increased lactic acid in the venous blood stimulates ventilation through receptors located in**

**A.** Carotid body
**B.** Medulla oblongata
**C.** Carotid sinus
**D.** Trachea

**Ans. 46. (A)**

The respiratory neurons are more responsive to $H^+$ ion concentration in the blood, but $H^+$ ion cannot cross the blood-brain barrier. But the peripheral chemoreceptor located in the carotid body can respond directly to the change in body $H^+$ ion concentration.

**47. A normal individual can expire out 80% of the air at the end of**

**A.** 1 sec **B.** 2 secs
**C.** 3 secs **D.** 5 secs

**Ans. 47. (A)**

Whent the forced vital capacity is timed it helps us to find out the amount of air expired out in unit time. As a standard the air expired at the end of 0.5 secs, 1 sec, 2 secs and 3 secs is measured. About 80% of the total vital capacity air should be expired out within 1st second and by the end of 3 second about 95% of the air should be expired out. It is abbreviated as $FEV_{0.5}$, $FEV_1$, $FEV_2$, $FEV_3$.

**48. Restrictive lung disease is associated with**

**A.** Normal $FEV_1$ / FVC ratio
**B.** Decreased $FEV_1$
**C.** Increased total lung capacity
**D.** Increased compliance of lung

**Ans. 48. (A)**

Restrictive disease is associated with decrease in the total lung capacity, but the amount of air that is expired out within 1st second is not decreased. The compliance is opposite to the elasticity. Fibrosis results in increase elasticity which decreases the compliance.

**49. Neural regulation of respiration is not affected by change in the**

**A.** Arterial carbon dioxide ion concentration
**B.** Arterial hydrogen ion concentration

C. Cerebrospinal carbon dioxide concentration
D. Cerebrospinal hydrogen ion concentration

**Ans. 49. (B)**

Though the primary stimulus to the chemosensitive area of the brain is hydrogen ion, increase in the $H^+$ ion concentration in the blood does not have a direct effect on the chemosensitive area because of blood-brain barrier which is impermeable to $H^+$ and bicarbonate ions. The CSF has very little protein buffer and hence $H^+$ ion concentration increases instantly when $CO_2$ enters CSF from the extensive subarachnoid plexus.

**50. Hypercapnoea is most likely to occur in**
A. Myasthenia gravis
B. Paralysis of the diaphragm
C. Depression of respiratory neurons
D. Cardiac asthma

**Ans. 50. (C)**

The level of the $CO_2$ in the body is regulated by the respiratory neurons and hence decreased activity of the respiratory neurons leads to accumulation of $CO_2$ in body. Myasthenia gravis and paralysis of diaphragm affects mostly the respiratory movement.

**51. At the normal arterial $PO_2$ of 95 mm Hg, the oxygen saturation of the arterial blood is**
A. 97% B. 40%
C. 100% D. 80%

**Ans. 51. (A)**

**52. With the decrease in the saturation of the hemoglobin with oxygen**
A. The content of $CO_2$ in the blood at a given $PCO_2$ decreases
B. Buffering capacity of hemoglobin decreases
C. Transport of $CO_2$ as carbamino compound increases
D. The $PCO_2$ for the given $CO_2$ content increases

**Ans. 52. (C)**

The capacity of hemoglobin to bind to carbon dioxide (carbamino compound) is more in the reduced form than the oxygenated form.

**53. When one-half the hemoglobin in blood gets bound to carbon monoxide**

**A.** At $PO_2$ of 100 mmHg, the percent saturation of Hb with $O_2$ will be below normal

**B.** At $PO_2$ of 50 mmHg, the percent saturation of Hb with $O_2$ will be below normal

**C.** At $PO_2$ of 100 mmHg, the dissolved form will be below normal

**D.** $P_{50}$ for oxygen will become less

**Ans. 53. (D)**

Binding of CO with Hb shifts the $O_2$ dissociation curve to the left. This decreases the partial pressure where available Hb is 50% saturated with $O_2$. Because CO has higher affinity for Hb it competes with $O_2$, the transport of $O_2$ becomes less.

**54. All of the following will increase the dead space volume EXCEPT**

**A.** Increase in the rate of breathing

**B.** Increase in the depth of breathing

**C.** Increased volume of anatomical dead space

**D.** Increased alveolar dead space

**Ans. 54. (A)**

In a normal quiet breathing with the tidal volume of 500 ml, 350 ml enters the alveoli and 150 remains in the dead space. Increase in the rate of respiration will increase the amount of air flowing through dead space per minute, but will not affect the amount air entering dead space per breath.

**55. Arterial hypoxia is usually associated with**

**A.** Increase in the $V_A/Q_c$ of the entire lung

**B.** Decreased $V_A/Q_c$

**C.** An higher arterial $PO_2$ than alveolar $PO_2$

**D.** Decreased shunting of blood from right to left side

**Ans. 55. (B)**

Decreased $V_A/Q_c$ (i.e. hypoventilation) is associated with increased arterial and alveolar $PCO_2$ and decreased arterial and alveolar $PO_2$.

**56. An increased difference between the partial pressure of alveolar oxygen and arterial oxygen is observed when**

**A.** Alveolar ventilation is reduced
**B.** Diffusing capacity decreases
**C.** $V_A/Q_c$ mismatched is increased
**D.** Ventilation and perfusion is doubled

**Ans. 56. (C)**

Decrease in alveolar ventilation will decrease $PO_2$ of alveoli as well as that of the arterial blood. Decrease in diffusing capacity only under extreme conditions will increase the difference.

**57. Which of the following statement about apneustic center is TRUE**
**A.** It is located in rostral pons
**B.** It initiates respiration
**C.** It can cause prolonged inspiration
**D.** It requires intact vagal innervation to cause apnoea

**Ans. 57. (C)**

Apneustic center is located in caudal pons. Apnoea means cessation of breathing, when apneustic center is stimulated it results in prolonged inspiration hence cessation of normal breathing cycle. Removal of pneumotaxic center and cutting of vagi is essential to produce apnoea.

**58. Prolongation of expiration is associated with**
**A.** Activity from slowly adapting stretch receptors
**B.** Activity from the apneustic center
**C.** Activity from rapidly adapting airway receptors
**D.** Decreased activity of the medullary expiratory neurons

**Ans. 58. (A)**

All others are associated with shortened expiratory period.

**59. The inability of the tissue to use oxygen results in**
**A.** Stagnant hypoxia
**B.** Histotoxic hypoxia
**C.** Hypoxic hypoxia
**D.** Anaemic hypoxia

**Ans. 59. (B)**

The stagnant hypoxia is due to the sluggish blood flow. Which results in decreased delivery of $O_2$ to the tissue.

In anaemic hypoxia the $O_2$ carrying capacity of the blood is less either due to decreased RBC count or presence of abnormal hemoglobin. Decreased $PO_2$ in the atmospheric air results in hypoxic hypoxia.

**60. The partial pressure of oxygen in the alveolar air is**

A. 40 mm Hg
B. 159 mm Hg
C. 104 mm Hg
D. 45 mm Hg

**Ans. 60. (C)**

Refer to Ans 37.

# 8

# Cardiovascular System

1. **Under normal physiological conditions, the strength of contraction of cardiac muscle fibre is controlled by**
   **A.** The number of fibres contracting
   **B.** Adjustment of intracellular contractile element
   **C.** ECF $Cl^-$ concentration
   **D.** The amplitude of the fast response AP

**Ans. 1. (B)**

According to Starling's law, the strength of contraction is directly proportional to the initial length of muscle fibres. At optimal length the contraction is maximum because of maximal interaction between actin and myosin.

2. **In cardiac muscle**
   **A.** Actin is absent in myofibrils
   **B.** AP may last several hundreds m secs
   **C.** Transverse tubules are absent
   **D.** ECF $Ca^{++}$ is not required for contraction

**Ans. 2. (B)**

The action potential in cardiac muscle is caused by two channels, Fast sodium channels and slow sodium calcium channels. These second set of channels are slow to open and remain open for a longer periods. Also the permeability of the cardiac membrane for the potassium decreases about 5 folds immediately after the onset of action potential. This prevent the outflux of potassium and this prevents the early return of action potential to the resting level.

3. **The all or none law, as applied to cardiac muscle states that all cardiac muscle cells**
   **A.** Are capable of generating AP
   **B.** Can be tetanised if the duration of AP is long

**C.** Are stimulated to contract if an AP is generated anywhere in the heart

**D.** Are refractory during the entire period of AP

**Ans. 3. (C)**

All the cardiac muscle cells are capable of generating an action potential. When one cell is stimulated it can excite the entire muscle mass because of it acts as an syncytium because of the cytoplasmic continuity.

**4. Cardiac glycosides such as digitalis enhance the performance of cardiac muscle cells by**

**A.** Inhibiting the Na K ATPase of cell membrane

**B.** Stimulating β adrenergic receptors

**C.** Stimulating intracellular production of cAMP

**D.** Blocking β adrenergic receptors

**Ans. 4. (A)**

Catecholamine acts by stimulation of beta-adrenergic stimulation. Digitalis inhibits sodium potassium ATPase, but increases the intracellular calçium which increases the contractility of the heart muscle.

**5. Positive inotropic agents**

**A.** Decrease maximum LV dp/dt

**B.** Shifts the LV curve to downward and to the right

**C.** Act through muscarinic receptors

**D.** Act through B adrenergic receptors

**Ans. 5. (D)**

Positive inotropic agents increase maximum left ventricular dp/dt and shifts the ventricular curve upward and to the left. Acetylcholine mediates it negative inotropic effect through muscarinic receptors.

**6. The delay of spread of impulse from atria to ventricle**

**A.** Aids in ventricular filling

**B.** Is because of small no. of large diameter fibres in AV node

**C.** Is recorded as QT interval in ECG

**D.** Is recorded as ST segment in ECG

**Ans. 6. (A)**

The delay in the spread of impulse in AV node helps atria to empty its blood in ventricle.

**7. The portion of the heart with the fastest conduction velocity for action potential is**

A. Left atrium B. Right atrium

C. Left ventricle D. Bundle of His

**Ans. 7. (D)**

Except the initial penetrating portion. The Bundle of His (AV bundle) has the characteristics opposite to that of AV node. It can transmit the impulse at the rate of 1.5 to 4 m/sec. This high conduction velocity in the Bundle of His and Purkinje's fibres is because of large number of gap junctions in the intercalated discs.

**8. Which of the following normally has a prominent prepotential**

A. SA node B. Atrial muscle

C. Bundle of His D. Ventricular muscle

**Ans. 8. (A)**

Prepotential is the change in membrane potential during the diastolic phase. This is because of the leaky sodium channels whose number is higher in SA node.

**9. The repolarisation of the ventricular muscle is caused by opening of**

A. Sodium channels B. Chloride channels

C. Potassium channels D. Calcium channels

**Ans. 9. (C)**

The depolarization in the ventricular muscle is caused by the fast sodium channels, while the plateau is because of slow sodium calcium channels. When the slow sodium calcium channels closes the only ions moving are potassium, which moves out of the cell resulting in the repolarisation of the membrane.

**10. Which of the following is incorrectly paired**

A. ACE: kinin metabolism

B. Vagal stimulation in neck: bradycardia

C. Prostaglandins: vasodilatation

D. Increased histamine: vasoconstriction

**Ans. 10. (D)**

Histamine is one of the most potent vasodilator. It is released during the allergic reactin which can lead to the anaphylactic shock. Vagal stimulation in the neck through baroreceptor brings about inhibition of vasomotor center.

**11. Vasopressin secretion is increased by**

A. Increased pressure in right ventricle
B. Increased pressure in aorta
C. Decreased pressure in right atrium
D. Increased pressure in right atrium

**Ans. 11. (C)**

Decreased pressure in the right atrium can be the result of increse in the negativity of the intrathoracic pressure or decreased venous return. Vasopressin is an antidiuretic hormone which tries to conserve the water in the body by decreasing the excretion of water through kidney. Increased pressure in the atrium stimulates the low volume- pressor area which causes the release of atrial natriuretic peptide.

**12. When a pheochromocytoma suddenly discharges a large amount of epinephrine into circulation, the heart rate would.**

A. Increase because of increased baroreceptor stimulation
B. Increase because epinephrine has a direct chromotropic effect
C. Increase because of increased blood pressure
D. Decrease due to increase vagal stimulation to heart

**Ans. 12. (B)**

Pheochromocytoma is the tumor of the adrenal medulla resulting in increased secretion of adrenalin. This increased adrenalin increases the blood pressure as well as the heart rate. Baroreceptor stimulation would decrease the heart rate, but by having direct effect on the heart epinephrine increases the heart rate.

**13. Blood pressure will show a rise in presence of**

A. Prostaglandins
B. ACE inhibitor
C. NO synthetase inhibitor
D. $V_1$ receptor inhibitor

**Ans. 13. (C)**

Prostaglandins are known to cause vasodilatation and hence decreases the blood pressure. Angiotensin converting enzyme (ACE) is required for conversion of angiotensin I to angiotensin II which is a powerful vasoconstrictor. NO (nitric oxide) is a vasodilator substance formed by the action of nitrogen synthetase.

**14. Angiotensin I brings about its action by activation of**

A. Guanylate cyclose
B. Protein kinase A
C. Phospholipase C
D. Phospholipase A2

**Ans. 14. (C)**

Phospholipase C is a membrane bound enzyme that splits the membrane lipid phosphoinositol biphosphate to inositol triphosphate and diacyl glycerol. This two than acts as second messenger.

**15. Endothelin I brings about its action by activation of**

A. Guanylate cyclose
B. Protein kinase A
C. Phospholipase C
D. Phospholipase A2

**Ans. 15. (D)**

Endothelin I is the most potent vasoconstrictor released by endothelim.

**16. Nitrous oxide brings about its action by activation of**

A. Guanylate cyclose
B. Protein kinase A
C. Phospholipase C
D. Phospholipase A2

**Ans. 16. (A)**

Nitrous oxide is a transmitter released synthesized from arginine by the enzyme no. synthetase. Guanylate cyclose is the membrane bound enzyme which when stimulated by 1st messanger coupled G protein causes the formation of cGMP from GTP.

**17. Noradrenalin brings about its action by activation of**

A. Guanylate cylase
B. Protein kinase A
C. Phospholipase C
D. Phospholipase $A_2$

**Ans. 17. (B)**

**18. The stage IV spontaneous depolarization in the pacemaker cells of SA node of heart**

A. Slows down with elevated ECF $K^+$
B. Results primarily from the decreased $K^+$ conductance

**C.** Results primarily from the decreased $Na^+$ conductance

**D.** Speeds up in the presence of acetylcholine

**Ans. 18. (B)**

Prepotential results from the overbalance of the sodium inflow to the potassium influx. This increases the resting membrane potential towards the positive side resulting in spontaneous depolarization. Acetylcholine increases the membrane permeability of the membrane for potassium, thus making the membrane potential more negative resulting in decreased excitability.

**19. The firing rate of cardiac pacemaker cell is primarily controlled by**

**A.** Duration of the refractory period

**B.** Threshold for slow channel activation

**C.** Rate of slow diastolic repolarisation

**D.** Magnitude of transmembrane potential

**Ans. 19. (C)**

Faster is the ascend of the diastolic potential faster the membrane will reach the threshold level, and will fire an action potential.

**20. Stimulation of the vagus nerve at the SA node results in**

**A.** Increased calcium influx

**B.** Increased potassium efflux

**C.** Decreased ECF potassium

**D.** Increased spontaneous depolarization

**Ans. 20. (B)**

**21. All of the following statement about cardiac cells are true EXCEPT**

**A.** 1st heart sound is caused by the closure of atrio-ventricular valve

**B.** Closure of atrioventricular valve causes C wave in the atrial pressure curve

**C.** V wave in the atrial pressure curve occurs before the opening of the atrioventricular valve

**D.** Atrial contraction is initiated by cells of the AV node

**Ans. 21. (D)**

Closure of the atrioventricular valve is the most important cause of the 1st heart sound. Closure of the atrioventricular valve pushes the atrioventricular ring in to the atrium which increases the pressure in the atrium (C wave). V wave of the atrial pressure is because of the rise in the intra-atrial pressure because of the venous return, which subsequently results in the opening of the atrioventricular.

**22. Maximum ventricular filling occurs during**

**A.** Isovolumic contraction period
**B.** Ejection period
**C.** Atrial diastole
**D.** Atrial systole

**Ans. 22. (C)**

Blood flows from the atria to the ventricle during atrial systole as well as atrial diastole. About 65–70% of the blood from the atria to the ventricle when both are in the state of diastole, the remaining 30–35% flows during atrial contraction.

**23. During isovolumic contraction period of the cardiac cycle**

**A.** Left atrial pressure falls slowly
**B.** Aortic pressure rises slowly
**C.** Left ventricular pressure rises rapidly
**D.** Aortic pressure rises rapidly

**Ans. 23. (C)**

During the isovolumic contraction period, the rate of rise of pressure in the left ventricle is maximum. The pressure rises from about 0–5 mmHg to 80 mmHg. During isovolumic contraction period the atrio-ventricular ring is pushed in the atria which results in rise in the atrial pressure. Isovolumic contraction period does not have any effect on aortic pressure.

**24. 2nd heart sound is associated with all of the following EXCEPT**

**A.** Aortic incisura
**B.** Onset of atrial contraction
**C.** Onset of IVR period
**D.** End of ventricular systole

**Ans. 24. (B)**

The aortic incisura is the notch produced due to the back flow of blood due to decease in the ventricular pressure with the beginning of ventricular diastole. This back flow results in the closure of the semilunar valve, producing 2nd heart sound.

**25. Which of the following statement is FALSE**

**A.** Rise in ventricular pressure is maximum during IVC

**B.** Fall in ventricular pressure is maximum during IVR

**C.** Semilunar valve closes at the end of proto-diastolic period

**D.** Maximum ventricular filling occurs during last rapid filling phase

**Ans. 25. (D)**

Last rapid filling phase of ventricle diastole coincides with the atrial systole. During this phase only 30–35% of the blood flows from atria to ventricles.

**26. In normal heart, current flows primarily from**

**A.** Apex to base of the ventricles

**B.** Epicardium to endocardium

**C.** Base to apex

**D.** None of the above

**Ans. 26. (A)**

**27. Each inch in horizontal direction in ECG paper is equal to**

**A.** 0.2 seconds **B.** 0.04 seconds

**C.** 1.0 seconds **D.** 0.8 seconds

**Ans. 27. (C)**

**28. When the distance between two successive R wave is 2 big square the heart rate would be**

**A.** 300 beats / min **B.** 150 beats / min

**C.** 100 beats / min **D.** 75 beats / min

**Ans. 28. (B)**

Heart rate can be calculated from ECG with the help of formula. HR = 300/ no. of big squares = 300/2 = 150 beats per minute, 1500/ no. of small squares.

**29. The cardiac electric axis of –40° is**

**A.** Called right axis deviation

**B.** Called left axis deviation

**C.** A normal axis

**D.** Seen in right vent. hypertrophy

**Ans. 29. (B)**

Although the mean electrical axis of the ventricles averages about 59°, the axis can swing to either the left to about 20 degrees or to the right to about 100 degrees even in a normal heart.

**30. Lead I measures the voltage difference between**

**A.** Right arm and left foot
**B.** Left arm and left foot
**C.** Left arm and right arm
**D.** Right foot and left foot

**Ans. 30. (C)**

Lead is the pair of the electrodes used to measure the voltage difference between two points. In lead one the positive terminal of the electrocardiograph is connected to the left arm and the negative terminal is connected to the right arm. Thus lead one measures the potential difference between left arm and right arm. Therefore when the point on the chest where the right arm connects to the chest is electronegative with respect to the point where the left arm connects the chest, the ECG records positively.

**31. Normal coronary blood flow to the heart is**

**A.** 70-80ml / 100gm
**B.** 7-8 ml / 100gm
**C.** 240 ml / 100gm
**D.** 30-50 ml / 100gm

**Ans. 31. (A)**

Heart receives blood supply through two coronaries. The heart receives 4-5% of the total cardiac output. Heart has a well developed autoregulatory mechanism.

**32. Dominance of coronary supply to the heart is dependent on the coronaries supplying the**

**A.** Base of the heart
**B.** Sinoatrial node
**C.** Crux of the heart
**D.** Atrioventricular node

**Ans. 32. (C)**

Crux is that part of the heart where all the grooves (i.e. interventricular and atrioventricular) meet. This area is present at the inferior portion of the heart.

**33. Normal electrical axis (vector) of ventricle is**

**A.** 100° **B.** 95°
**C.** 59° **D.** 90°

**Ans. 33. (C)**

Electrical axis or vector is the average direction of the current flow during the process of the depolarization in the ventricle. The normal current flow direction is from apex to base in the 59° direction.

34. **Bipolar limb leads helps in the diagnosis of**
    **A.** Diseases of interventricular septum
    **B.** Diseases of left ventricular myocardium
    **C.** Arrhythmias of heart
    **D.** All of the above

**Ans. 34. (C)**

The recordings in the bipolar limb leads are similar to one another. To diagnose one needs to find out the time relation between to waves, which can be easily done in bipolar limb leads. When one want to diagnose the disease of the atrial or ventricular muscle, conducting system it does matter which leads are selected as some leads might be affected while some might show normal recordings.

35. **Infarction of inferior wall of heart can be diagnosed by studying**
    **A.** II, III, aVF limb leads
    **B.** I, aVL limb lead
    **C.** aVR, aVL, aVF limb leads
    **D.** I, aVR limb leads

**Ans. 35. (A)**

The lead II, III and aVF are so placed that they look at the heart inferiorly, this helps these leads to record events in the inferior wall.

36. **The P wave of ECG denotes the electrical activity of**
    **A.** AV node **B.** SA node
    **C.** Purkinje fibres **D.** Ventricles

**Ans. 36. (B)**

P wave is the first positive deflection in the ECG recording. It is caused due to atrial depolarization. The P wave is followed by atrial systole.

37. **Atrial depolarization in a normal ECG is denoted by**
    **A.** P wave **B.** QRS complex
    **C.** T wave **D.** U wave

**Ans. 37. (A)**

**38. Which of the following statement about ECG is FALSE**

**A.** Atrial repolarisation wave is obscured by QRS complex

**B.** First wave of ventricular depolarization is negative

**C.** Ventricular repolarisation wave is normally negative

**D.** RR interval can be used to determine the heart rate

**Ans. 38. (C)**

Atrial repolarisation wave is obscured by the electrical events in the ventricle. First wave of the ventricular depolarization is Q wave which is negative. RR interval is the distance between two consecutive R waves, of consecutive heart beat.

**39. Normal duration of PR interval is**

**A.** 0.10 secs **B.** 0.20 secs

**C.** 0.16 secs **D.** 0.22 secs

**Ans. 39. (C)**

PR interval signifies the time taken for an impulse to travel from atria to ventricle. It is normally 0.16 seconds. Though interval more than 0.20 seconds is taken as abnormal. PR interval is prolonged in any condition which would decrease rate of transmission from SA node to ventricle. It is shortened in Wolff-Parkinson-White syndrom.

**40. Incisura in the aortic pressure curve is**

**A.** Because of increased aortic pressure

**B.** Associated with third heart sound

**C.** Because of running away of blood to the periphery

**D.** Because of backward flow of blood from aorta to ventricle

**Ans. 40. (D)**

**41. Which of the following is TRUE**

**A.** Increase heart rate increases the duration of cardiac cycle

**B.** 1st heart sound marks the end of the isovolumic contraction period

C. Maximum flow from atria to ventricle occurs during diastole

D. 2nd heart sound coincides with carotid pulse

**Ans. 41. (C)**

Heart rate and cardiac cycle duration are inversely related. Thus increase in heart rate decreases the duration of cardiac cycle. 1st heart sound marks the beginning of ventricular systole, the 1st heart sound coincides with carotid pulse.

**42. The isovolumic contraction period in the cardiac cycle of duration 0.8 sec, would last for**

**A.** 0.14 sec | **B.** 0.08 sec
**C.** 0.04 sec | **D.** 0.05 sec

**Ans. 42. (D)**

The isovolumic contraction period is the first period of the ventricular systole. Out of the total duration of 0.3 seconds IVC period lasts for 0.05 seconds.

**43. Considering cardiac cycle duration to be 0.8 sec, the protodiastolic period would last for**

**A.** 0.14 sec | **B.** 0.08 sec
**C.** 0.04 sec | **D.** 0.10 sec

**Ans. 43. (C)**

The protodiastolic period is the first period of the ventricular diastole. It is the period between beginning of ventricle diastole and closure of the semilunar valves.

**44. Considering cardiac cycle duration to be 0.8 sec, the isovolumic relaxation period would last for**

**A.** 0.14 sec | **B.** 0.10sec
**C.** 0.04 sec | **D.** 0.05 sec

**Ans. 44. (A)**

**45. Considering cardiac cycle duration to be 0.8 seconds, the reduced ejection period would last for**

**A.** 0.14 sec | **B.** 0.08 sec
**C.** 0.10 sec | **D.** 0.05 sec

**Ans. 45. (B)**

The ejection phase of the ventricle lasts for 0.25 seconds. During the initial part because of greater pressure difference between ventricles and arteries (aorta and pulmonary trunk) and because of greater volume of the blood in the ventricle the blood flowing from ventricles to arteries is greater and during the later half the ejection becomes less.

**46. In the heart dipoles are formed when**

**A.** Only a part of the heart is depolarized
**B.** Complete heart is depolarized
**C.** Complete heart is repolarized
**D.** There is ischemic heart disease

**Ans. 46. (A)**

Dipoles are the pair of opposite charges lying close to each other. This happens when only a part the heart is depolarized. The depolarized part has the negativity outside while the adjacent polarized part has the positivity outside constituting a Dipole.

**47. All of the following statements about dipole are true EXCEPT**

**A.** Dipole results in the flow of the current along its axis
**B.** Multiple dipoles in the same plane results in the large resultant vector
**C.** A dipole generates large field around it
**D.** Current traveling perpendicular to the axis of the dipole is picked by the electrodes

**Ans. 47. (D)**

Dipole results in a flow of the current from negative charges to the positive charges. This generates large field around a dipole. The current traveling in a straight line (along the axis) is the one picked by the electrodes.

**48. A record in the electrocardiogram is positive**

**A.** When the apex is depolarized
**B.** When the base is depolarized
**C.** When the current flows from the base to apex
**D.** When the cardiac vector moves toward an active electrode

**Ans. 48. (D)**

**49. The conduction velocity of the fibres in the atrioventricular node is**

**A.** 0.5 m/s
**B.** 0.05 m/s
**C.** 5 m/s
**D.** 0.005 m/s

**Ans. 49. (B)**

**50. AV nodal delay is because of**

A. Less number of fibres
B. Less number of voltage gated sodium channels
C. Increase glycogen content
D. More negative RMP

**Ans. 50. (D)**

The resting membrane potential of AV nodal fibres is more negative about –85 to –95 mv, hence it takes longer time for the AV nodal fibres to reach the firing level.

**51. AV nodal delay is**

A. Lengthened by sympathetic stimulation
B. Shortened by sympathetic stimulation
C. Shortened by vagal stimulation
D. Is not affected by autonomic nervous response

**Ans. 51. (B)**

**52. All of the following statements about Purkinje's fibres are true EXCEPT**

A. It excites only those cardiac muscles located near endocardial surface
B. It conducts impulse at a rate of 1 to 4 m/s
C. It has few gap junctions
D. It allows simultaneous contraction of both the ventricles

**Ans. 52. (C)**

Purkinje's fibres have many gap junctions which allows the rapid flow of current and hence the conduction velocity is maximum in Purkinje's fibres. This rapid conduction allows simultaneous contraction of both the ventricles.

**53. Which of the following is last to be depolarized**

A. Apex of the left ventricle
B. Endocardium of the left ventricle
C. Base of the left ventricle
D. Lower most portion of the interventricular septum

**Ans. 53. (C)**

The last portion of the heart to be depolarized is postero basal part of left ventricle, pulmonary conus, uppermost part of the interventricular septum.

**54. The 1 millimetre deflection in vertical direction indicates the current input of**

A. 0.1 millivolts B. 1 millivolts
C. 0.04 millivolts D. 2 millivolts

**Ans. 54. (A)**

The ECG paper is calibrated such that 10 divisions upward or downward represent 1 millivolt. Each division is 1 millimeter.

**55. One millimeter horizontal distance in ECG paper is equivalent to**

A. 4 seconds B. 0.04 seconds
C. 0.04 milliseconds D. 1 second

**Ans. 55. (B)**

One big square in horizontal direction is one inch. This big square is divided in to 5 small square, thus each small square is 5 mm. This small square is further divided in to smaller 5 squares, thus each smallest square is 1 mm. So the 1 second is subdivided in to 25 small square each smallest square thus measures 0.04 seconds.

**56. The time taken for impulse to travel to travel from SA node to ventricle is represented by**

A. P–R interval B. S–T segment
C. Q–T segment D. T–P segment

**Ans. 56. (A)**

P wave represents the atrial repolarisation wave and electrical activity in SA node. R wave represents the ventricular repolarisation wave.

**57. ECG changes in myocardial infarction results due to**

A. Rapid depolarization
B. Delayed repolarisation
C. Increase resting membrane potential
D. None of the above

**Ans. 57. (D)**

ECG changes in MI results due to delayed depolarization (which occurs ½ hour after infarction), rapid repolarisation, (which occurs within seconds of infarction) and decrease RMP (which occurs minutes after infarction).

**58. Splitting of second heart sound**

A. Is due to early closure of pulmonary valve
B. Widens during deep inspiration

**C.** Widens during deep expiration
**D.** Is seen in aortic stenosis

**Ans. 58. (B)**

Splitting of second heart sound is due to early closure of aortic valve because of greater pressure difference. During deep inspiration there is increased venous return which increases the blood flow to right ventricle, thus pulmonary valve remains open for a longer time. Aortic stenosis results in reverse splitting of 2nd heart sound.

**59. Ventricular diastole**

**A.** Follows QRS complex
**B.** Begins just prior to the closure of the semilunar valves
**C.** Decreases the coronary circulation
**D.** Results in C wave in the atrial pressure curve

**Ans. 59. (B)**

Ventricular diastole follows the T wave of ECG. Ventricular diastole leads to the fall in the pressure in the ventricles which results in the back flow of the blood from aorta and pulmonary trunk to LV and RV respectively. This leads to the closure of the semilunar valves.

**60. The conduction velocity in sinoatrial node fibres is**

**A.** 0.05 meter/second
**B.** 0.5 meter/second
**C.** 1 meter/second
**D.** 2 meter/second

**Ans. 60. (A)**

The conduction velocity in the (a) SA and AV node is 0.05 m/sec (according to Ganong), (b) atrial and ventricular muscle fibres is 0.3–0.5 m/sec (according to Guyton, but according to Ganong conduction velocity in ventricular muscle fibre is 1 m/sec), (c) internodal tract is 1 m/sec.

**61. The conduction velocity in atrial muscle fibres is**

**A.** 0.5 meter/second **B.** 0.5 meter/second
**C.** 1 meter/second **D.** 2 meter/second

**Ans. 61. (A)**

See answer 60

**62. The conduction velocity in ventricular muscle fibres is**

**A.** 0.5 meter/second **B.** 0.5 meter/second
**C.** 0.05 meter/second **D.** 2 meter/second

**Ans. 62. (A)**

See answer 60

**63. The conduction velocity in internodal bundles fibres is**

**A.** 0.05 meter/second **B.** 0.5 meter/second
**C.** 1 meter/second **D.** 2 meter/second

**Ans. 63. (C)**

See answer 60

**64. Considering cardiac cycle duration to be 0.8 sec, the atrial systole would last for**

**A.** 0.14 sec **B.** 0.05sec
**C.** 0.10 sec **D.** 0.05 sec

**Ans. 64. (C)**

**65. Which of the following is NOT a property of a cardiac muscle**

**A.** Excitability
**B.** Conductivity
**C.** Rhythmicity
**D.** Short refractory period

**Ans. 65. (D)**

Heart is one of the few tissue in the body which shows rhythmicity. It is also capable of conducting the impulse. It has a long refractory period.

**66. Tetanisation of the heart is prevented by its property of**

**A.** Rhythmicity
**B.** Long refractory period
**C.** Conductivity
**D.** Excitability

**Ans. 66. (B)**

Refractory period is a period during which the cell cannot be excited again. The absolute refractory of cardiac muscle is very long and hence however strong the stimuli the muscle does not respond again. This prevents the heart from being tetanised.

**67. The fast response action potential is seen in**
A. Sinoatrial node
B. Atrioventricular node
C. Ventricular muscle
D. None of the above

**Ans. 67. (C)**

The action potential in the cardiac tissue is of two types, depending upon the channels present in the cell membrane. Conducting tissues such as SA node and AV node have many leaky sodium channels and slow sodium-calcium channels. The depolarization phase of these is because of the opening of the slow sodium-calcium channel, thus the ascend is slow and gradual, this is the slow action potential. In Purkinje's fibres and ventricular muscle fibres the action potential is because of the fast sodium channels, thus the response is fast.

**68. An ability of a tissue to respond to a stimulus is called**
A. Excitability
B. Distensibility
C. Conductivity
D. Rhythmicity

**Ans. 68. (A)**

The distensibility is the ability to increase the diameter in response to the stretch. Conductivity is the ability to allow the impulse to pass through and to carry it from one point another. Rhythmicity is the ability to generate impulse on its own.

**69. The impulse from sinoatrial node reaches the ventricles by**
A. 0.22 seconds
B. 0.16 seconds
C. 0.8 seconds
D. 0.03 seconds

**Ans. 69. (A)**

The pulse takes about 0.22 seconds to pass through the various conducting tissue and reach the epicardium of the ventricles. Since Purkinje's fibres conduct impulse at a very fast rate, both the ventricles contract simultaneously.

**70. Which of the following sentence is true**
A. Conduction velocity is directly proportional to the glycogen content of the fibre

B. Rhythmicity is directly proportional to the glycogen content of the fibre
C. Sinoatrial node fibre has the highest glycogen content
D. Duration of systole is directly proportional to the glycogen content

**Ans. 70. (A)**

Conduction velocity is of the tissue is directly proportional to its glycogen content, while the rhythmicity and the duration of systole is inversely proportional to its glycogen content. SA has the lowest glycogen and maximum autorhythmicity.

**71. The pacemaker activity of the sinoatrial node is increased by all of the following EXCEPT**

A. Epinephrine
B. Increased basal metabolic rate
C. Stimulation of the right vagus nerve
D. Throxine

**Ans. 71. (C)**

**72. All of the following statements about second heart sound are true EXCEPT**

A. It is short, sharp and high pitched
B. It indicates end of the ventricular systole
C. It coincides with the spike of the R wave of the ECG
D. It occurs just after carotid pulsations

**Ans. 72. (C)**

R wave is the wave of ventricular depolarization, which precedes ventricular systole and hence before 2nd heart sound.

**73. Which of the following statement about first heart sound is FALSE**

A. It is dull, prolonged and low pitched
B. It is heard best over the apex area
C. It indicates the end of clinical ventricular systole
D. Its intensity indicates the condition of the myocardium

**Ans. 73. (C)**

1st heart sound is because of the closure of the atrioventricular valve. Closure of the valve results at the end of the atrial contraction. Atrial systole is followed by ventricular systole which results in increase in intraventricular pressure and thus the blood tries to go back to atria resulting in the closure of the atrioventricular valve.

**74. Sinoatrial node is situated in**

**A.** Right ventricle **B.** Right atrium
**C.** Left atrium **D.** Left ventricle

**Ans. 74. (B)**

The sinoatrial node is the pacemaker of the heart. It is situated in the superior part of the right atrium next to the opening of the superior vena cava.

**75. Atrioventricular node is located in**

**A.** In right atrium behind tricuspid valve
**B.** In the posterior septal wall of left atrium
**C.** Near the opening of the superior vena cava
**D.** It the upper part of interventricular septum

**Ans. 75. (A)**

Atrioventricular node is situated next to the opening of the coronary sinus in the posterior septal part of the right ventricle.

**76. Atrioventricular nodal delay is caused due to all of the following EXCEPT**

**A.** More negative resting membrane potential of their fibres
**B.** Smaller size of their fibres
**C.** Few gap junctions in their fibre membrane
**D.** High glycogen content of their fibres

**Ans. 76. (D)**

Atrioventricular node allows atria to empty its blood completely in to the ventricle. The conduction velocity is inversely proportional to the glycogen content of the fibres.

**77. The end diastolic volume of the ventricle increases when**

**A.** The person is upright
**B.** There is increase in the venous tone
**C.** There increase in intrapericardial pressure
**D.** There is decrease in ventricular compliance

**Ans. 77. (B)**

Increase in the tone of the veins results in increase in the venous return to the heart. Increase in the venous return leads to increase flow of the blood from the atria to the ventricles.

**78. Which of the following would decrease the end diastolic volume of the ventricles**

A. Increased blood volume

B. increased pumping action of the skeletal muscle

C. Positive intrathoracic pressure

D. Stronger atrial contractions

**Ans. 78. (C)**

Increase in the blood volume increases the mean circulatory filling pressure which increases the venous return. Increase in the pumping action of the skeletal muscle increases the return of the blood from the periphery to the heart. Increased positive intrathoracic pressure decreases the venous return and prevents the full dilatation of the heart during diastole.

**79. The beating heart at rest consumes**

A. 7-9 ml of oxygen per minute

B. 2-4 ml of oxygen per minute

C. 0.7-0.9 ml of oxygen per minute

D. 17-19 ml of oxygen per minute

**Ans. 79. (A)**

Heart has one of the highest oxygen extraction ratio. When the activity of the heart is artificially reduced as during cardiovascular operation the oxygen consumption falls to 2-4 ml of oxygen per minute.

**80. The oxygen consumption by the heart is increased by all of the following EXCEPT**

A. Increased myocardial tension

B. Decreased afterload

C. Increased contractility of the myocardium

D. Increased heart rate

**Ans. 80. (B)**

Afterload is the pressure (load, resistance) against which the heart has to pump the blood. So increase in afterload would result in extra pressure on the myocardium. All the other factor increases the oxygen consumption by myocardium due to increased activity.

**81. Cardiac output will be normal or slightly increased in**

A. Haemorrhagic shock

B. Anaphylactic shock

C. Septicemic shock
D. Cardiogenic shock

**Ans. 81. (C)**

Cardiac output is dependent on the metabolic needs of the body. In septicemia the metabolic activity of the cells of the body increases and to meet this demand the output increases. But when even this increased cardiac output is not sufficient to meet the needs of the body a shock results due to mismatched demand and the supply. Hemorrhagic shock results from the trauma or the internal bleeding or the loss of body fluids. In anaphylactic shock there is excessive vasodilatation resulting in accumulation of the blood in veins resulting in decreased venous return and hence cardiac output. In cardiogenic shock the ability of the heart to pump the blood is decreased.

**82. Cardiogenic shock can result due to**

A. Decreased venomotor tone
B. Decreased blood volume
C. Severe valvular dysfunctions
D. Increased metabolism of body

**Ans. 82. (C)**

Cardiogenic shock results from decreased ability of the heart to pump the blood it receives. Severe valvular dysfunction either acts as a block for the passage of the blood (as in stenosis) or a passage for the blood to flow backwards, both of which results in decreased cardiac output.

**83. The cause for the progression of the shock to the irreversible stage is**

A. Decreased ATP concentration of the cell
B. Electrolyte imbalance
C. Decreased temperature of the body
D. Release of the toxins

**Ans. 83. (A)**

For a cell to perform the function it requires energy which is provided by the ATP. In shock when a cell does not receive enough nutrients, oxygen and blood, the used ATP molecule gets converted to adenosine (ATP - ADP - AMP - Adenosine). Once adenosine leaves the cell it cant be utilized to form ATP. Thus the enegy supply further decreases.

**84. Adrenalin would be the drug of choice in**

**A.** Anaphylactic shock
**B.** Traumatic shock
**C.** Septicemic shock
**D.** All of the above

**Ans. 84. (A)**

Anaphylactic shock results due to allergic reaction. The allergic reaction results in the release of histamine which is a powerful vasodilator, thus there is decreased vascular tone and the blood accumulates in the veins. Adrenalin being a vasoconstrictor and cardiotonic agent helps to restore the vascular tone Thus helps to increase the venous return cardiac output.

**85. Vasovagal syncope is an example of**

**A.** Neurogenic shock
**B.** Hypovolumic shock
**C.** Septicemic shock
**D.** Cardiogenic shock

**Ans. 85. (A)**

Neurogenic shock results from decreased vasomotor activity or increased parasympathetic discharge. Vasovagal syncope results when there is a strong emotional reaction. This results in stimulation of parasympathetic nervous system. Stimulation of parasympathetic nerves decreases the activity of the heart and peripheral resistance.

# 9

# Central Nervous System (CNS)

1. **Commonest type of the synapse found in human is**
   A. Axosomatic synapse
   B. Axodendritic synapse
   C. Axoaxonic synapse
   D. Dendrodendritic synapse

**Ans. 1. (B)**

Synapse is the junction between two neurons. Most of the synapse is formed by axons of presynaptic neuron with dendrites of postsynaptic neurons (80–95%), remaining is axosomatic (5–20%), and small part is constituted by axoaxonic (for presynaptic inhibition) and dendrodendritic.

2. **The quantity of transmitter released by presynaptic terminal in to the synaptic cleft depends on**
   A. RMP of the nerve
   B. Amount of calcium entering the presynaptic terminal
   C. Amount of mitochondria in presynaptic terminal
   D. Number of postsynaptic receptors

**Ans. 2. (B)**

The presynaptic membrane contains many voltage gated calcium channels which opens with the arrival of an action potential. This leads to influx of calcium. This calcium binds with the proteins present on the inner surface of the presynaptic membrane (release sites), this causes binding of the transmitter vesicles with the presynaptic membrane then by the process of exocytosis the transmitter is released in to the synaptic cleft.

**3. The G proteins present in the neuronal cell membrane are a complex of**

**A.** α, β, γ **B.** α, β, δ
**C.** β, δ, γ **D.** β, γ, δ

**Ans. 3. (A)**

G proteins are the membrane regulatory proteins which many 1st messenger bind with. This G protein then either activates a membrane bound enzyme or cytoplasmic enzymes leading to the formation of the second messenger. The G protein itself is a complex of α, β and γ subunits.

**4. Which of them following is a slow acting neurotransmitter**

**A.** Acetylcholine **B.** Dopamine
**C.** Nitric oxide **D.** β endorphin

**Ans. 4. (D)**

All others are the small-molecule, rapidly acting neurotransmitter. β-endorphin is a neuropeptide (pituitary peptide) which is slow acting. The neuropeptides are synthesized as integral part of the large protein molecules by the ribosomes in the neuronal cell body. Then this transmitter splits from the protein molecules and are packed in to the vesicles by Golgi complex and are then carried to the tip of the nerve fibres by axial streaming of the axon cytoplasm. Once relasead in to the cleft they are slow to act but their action is prolonged.

**5. Which of them following is a rapid acting neurotransmitter**

**A.** TRH **B.** Enkephalin
**C.** Histamine **D.** Bradykinin

**Ans. 5. (C)**

Rest all are neuropeptides which are slow to act. Histamine belongs to Class II (amines) of the small-molecule rapidly acting neurotransmitters. These small-molecule rapidly acting neurotransmitters are synthesized in the cytoplasm of the presynaptic membrane and then are actively transported in to the vesicles and then are released when an action potential arrives. Once released it brings about it action on the postsynaptic neuronal receptor within few milliseconds.

**6. Which of them following is not a slow acting neurotransmitter**

A. Prolactin B. Sleep peptides
C. Glutamate D. Substance P

**Ans. 6. (C)**

Glutamate is a rapidly acting transmitter belonging to Class III (amino acids) of small- molecule rapidly acting neurotransmitters. Rest all are slow acting.

**7. Which of them following is a not a rapid acting neurotransmitter**

A. Norepinephrine B. Serotonin
C. GABA D. Angiotensin II

**Ans. 7. (D)**

**8. Which of the following neurotransmitter is secreted by median raphe of brain**

A. Acetylcholine B. Serotonin
C. GABA D. Nitric oxide

**Ans. 8. (B)**

Serotonin is a (amine, class II) small- molecule rapidly acting neurotransmitters. It is secreted by nuclei that originate in the median raphe of the brainstem and project to many brain and spinal cord areas, especially to the dorsal horns of the spinal cord and to the hypothalamus. It acts as an inhibitor of pain pathways and as mood controller.

**9. Acetylcholine is secreted by**

A. Pyramidal cells of motor cortex
B. Locus ceruleus
C. Synapses in spinal cord
D. Cerebellum

**Ans. 9. (A)**

Acetylcholine is secreted by many areas of the brain but especially by large pyramidal cells of the motor cortex, different neurons in the basal ganglia,motor neurons innervating skeletal M, preganglionic of ANS, post-ganglionic parasympathetic neurons.

**10. Which of the following neurotransmitter is secreted by substantia nigra of brain**

A. Acetylcholine B. Serotonin
C. GABA D. Dopamine

**Ans. 10. (D)**

The dopaminergic neurons originating in substantia nigra terminates in striatal region of basal ganglia. It is usually a inhibitory neurotransmitter.

**11. Norepinephrine is secreted by**

**A.** Pyramidal cells of motor cortex
**B.** Locus ceruleus
**C.** Synapses in spinal cord
**D.** Basal ganglia

**Ans. 11. (B)**

Norepinephrine is secreted by many neurons in the brainstem (locus ceruleus in pons) and hypothalamus. From here the fibres extend to the wide areas of the brain and helps to control the overall activity and mood of the mind. It is mostly an excitatory neuron but also has inhibitory effects.

**12. Glycine is secreted by**

**A.** Synapses in spinal cord
**B.** Median raphe
**C.** Basal ganglia
**D.** Cerebellum

**Ans. 12. (A)**

Glycine mostly has inhibitory effects.

**13. Nitric oxide**

**A.** Is preformed and stored in vesicles
**B.** Is released by exocytosis
**C.** Is synthesized instantly as needed
**D.** Decreases RMP in postsynaptic neurons

**Ans. 13. (C)**

Nitric oxide is found in the areas of the brain concerned with learning and behaviour. It is not preformed in the synaptic membrane but is synthesized as and when need arises. It then affects the metabolic activity of the cell without affecting the membrane potential.

**14. Which of the following neurotransmitter is involved in long-term behaviour and memory**

**A.** Acetylcholine **B.** Norepinephrine
**C.** GABA **D.** Nitric oxide

**Ans. 14. (D)**

**15. Which of the following statement about neuropeptide is FALSE**

**A.** It is released in large amount
**B.** Is less potent than small molecule transmitters

C. Its action is short lived

D. It is synthesized in the cytosol of the presynaptic membrane

**Ans. 15. (C)**

The process of the formation of neuropeptide is very laborious. Thus it is released in less amount but is makes for this small amount by having a more prolonged and more potent effect.

**16. Presynaptic terminals of sensory pathways secret**

A. Acetylcholine B. Dopamine
C. Glutamate D. Glycine

**Ans. 16. (C)**

Glutamate is predominantly an excitatory neurotransmitter released by sensory tracts and some areas of the cortex.

**17. Soma of the motor neurone has the RMP of**

A. – 65 mv B. – 90 mv
C. – 105 mv D. – 80 mv

**Ans. 17. (A)**

This lower voltage allows the positive as well as negative control of the excitability of the neuron. This is the basis for the two mode of the functions of the neurons.

**18. Action potential in a neuron begins in**

A. Soma B. Dendrite
C. Terminal buttons D. Axon hillock

**Ans. 18. (D)**

The RMP of the soma is in large part due to the presence of relatively high ICF potassium concentration. Also the somal membrane has very few sodium channels and thus it fails to fire an action potential. The initial segment (axon hillock) has many sodium channels and thus is capable of generating an action potential.

**19. The short circuiting mechanism for inhibiting neuron is because of**

A. $Na^+$ channels B. $Cl^-$ channels
C. $K^+$ channels D. $Ca^{++}$channels

**Ans. 19. (B)**

A tendency of the chloride channels to maintain membrane potential near the resting value when the when the inhibitory channels are wide open is called 'short circuiting of the membrane'. In some neurons the concentration difference for the chloride across the membrane causes chloride Nernst potential that is exactly equal to its resting potential. Thus there is no net flow of chloride ions to cause an inhibitory postsynaptic potential. With excitatory signal and opening of the sodium channel the Nernst potential for chloride is disturbed and along with sodium, chloride also moves. Thus for the mem-brane to be excited extra sodium influx is required.

**20. The presynaptic inhibition is caused by**

**A.** Glycine  **B.** Glutamate
**C.** GABA  **D.** Norepinephrine

**Ans. 20. (C)**

GABA causes opening of the anion (chlorine) channels, which nullifies the change in potential because of sodium influx. Presynaptic inhibition occurs in many of the sensory pathways, which minimizes the sideways spread of signals in the sensory tracts.

**21. The seat of fatigue in the intact human body is**

**A.** Neuromuscular junction
**B.** Peripheral synapse
**C.** Central synapse
**D.** Muscle fibres

**Ans. 21. (C)**

The seat of fatigue in a nerve - muscle preparation is a myoneural junction, but in the inside the body it is the synapses present in the brain which are the sites of fatigue. This helps to prevent any chances of injury to the muscle.

**22. Which of the following decreases the synaptic excitability**

**A.** Caffeine  **B.** Alkalosis
**C.** Cocoa  **D.** Anaesthetic agent

**Ans. 22. (D)**

All other are the stimulants of the synaptic activity. Alkalosis increases the synaptic activity by decreasing the ionic portion of the calcium.

**23. The synaptic activity is increased by**

**A.** Acidosis  **B.** Tea
**C.** Hypoxia  **D.** Anaesthetic agent

**Ans. 23. (B)**

**24. The minimum duration of synaptic delay is**

A. 0.5 millisecond
B. 0.05 millisecond
C. 0.005 millisecond
D. 5 millisecond

**Ans. 24. (A)**

A delay occurs in the transmission of an action potential from presynaptic to postsynaptic membrane. This delay is called a synaptic delay and is due to 1. release of neurotransmitter from presynaptic neurone 2. diffusion of the neurotransmitter to postsynaptic neurone 3. action of a neurotransmitter on postsynaptic membrane receptors 4. change in the membrane permeability 5. increase or decrease flux of ions. By noting the time interval between input and output valley number of neurons can be estimated.

**25. A protection against excessive neuronal activity is because of**

A. Synaptic fatigue
B. Synaptic plasticity
C. Presynaptic inhibition
D. Synaptic delay

**Ans. 25. (A)**

Synaptic fatigue is the absence of the postsynaptic neurons response to presynaptic stimulus. This results from the exhaution of the neurotransmitter.

**26. All of the following are the property of the synapse EXCEPT**

A. Synaptic delay
B. Post-tetanic facilitation
C. Two way conduction
D. Summation

**Ans. 26. (C)**

Synaptic transmission involves release of neurotransmitter and its action on the receptors on the postsynaptic membrane. Synaptic conduction is always one way.

**27. Synaptic fatigue results due to all of the following EXCEPT**

A. Exhaustion of neurotransmitter
B. Progressive inactivity of receptors in postsynaptic membrane
C. Build up of calcium in presynaptic neuron
D. Slow build up of calcium in postsynaptic neurons

**Ans. 27. (C)**

Slow build up of calcium in the postsynaptic neuron causes opening of calcium activated potassium channels and outflux of potassium and change in the membrane potential. Build up of $Ca^{++}$ in presynaptic neurons leads to post-tetanic facilitation.

**28. An increased synaptic response to the subsequent stimuli following a rapid repetitive stimuli is due to the property of**

**A.** Recruitment
**B.** Post-tetanic facilitation
**C.** Synaptic delay
**D.** Convergence

**Ans. 28. (B)**

Rapid repetitive stimuli results in the excessive build up of calcium in the presynaptic membrane (because of the slowness of the calcium pump), causing release of more neurotransmitter. The post-tetanic facilitation is thought to be helpful in the short-term memory.

**29. Strychnine increases the synaptic excitability by**

**A.** Lowering the excitatory threshold of the neurons
**B.** Increasing the number of the receptors on the postsynaptic membrane
**C.** Decreasing the reuptake of neurotransmitter
**D.** Inhibiting the action of inhibitory neurotransmitter

**Ans. 29. (D)**

Strychnine is especially known to inhibit the effect of the glycine in the spinal cord.

**30. An ability of the postsynaptic neuron to fire an action potential when many presynaptic nerve terminals are stimulated together is called**

**A.** Temporal summation
**B.** Post-tetanic facilitation
**C.** Spatial summation
**D.** Recruitment

**Ans. 30. (C)**

[It is known that change in the potential at any point in the soma (body) of the neuron will cause the similar potential changes everywhere in the soma. That is why with excitation of excitatory synapses there effect on the soma will be added up and if these changes the potential to a threshold level an action potentials will be fired. Temporal summation is time related.

**31. The specificity of a nerve fibre for the different sensation is determined by [Labelled line principle]**

**A.** The type of a nerve fibre
**B.** The diameter of the nerve fibre
**C.** RMP of the nerve fibre
**D.** Site of termination of nerve fibre in the brain

**Ans. 31. (D)**

**32. The rise of receptor potential above the threshold level is associated with**

**A.** Increase in the magnitude of the AP
**B.** Increase in frequency of AP
**C.** Increase in receptor fatigability
**D.** Decrease in frequency of AP

**Ans. 32. (B)**

Action potential follows all or none phenomenon, i.e. when the membrane potential reaches the threshold level action potential will result with the maximum amplitude. So the rise of receptor potential will not change the magnitude of an action potential but if the receptor potential rises above the threshold level more and more action potential will be fired.

**33. Majority of the peripheral sensory and post-ganglionic nerve fibre belongs to**

**A.** Type A β fibres **B.** Type A δ fibres
**C.** Type C fibres **D.** Type B fibres

**Ans. 33. (C)**

**34. The adaptation in the pacinian corpuscles results from**

**A.** Redistribution of the fluid within the corpuscles
**B.** Accommodation due to closure of the sodium channels
**C.** Both A and B are correct
**D.** None of the above

**Ans. 34. (C)**

Pacinian corpuscles are responsible for the detection of pressure. It is fast adapting receptor. Pacinian corpuscle is a viscoelastic structure and hence sudden force would result in the distortion and this will be transmitted through the viscous component to the central nerve terminal. After a while the force will be distributed equally to the entire corpuscle. Along with this the sodium channels at the tip of the nerve terminal closes.

**35. The sensation of fine touch from the lower limb is carried by**

**A.** Fasciculus gracilis
**B.** Fasciculus cuneatus
**C.** Anterior spinothalamic tract
**D.** Lateral spinothalamic tract

**Ans. 35. (A)**

Fasciculus gracilis and fasciculus cuneatus forms the posterior column tract (dorsal leminiscal system). These are composed of large myelinated nerve fibres and transmit the impulses at a velocity of 30–110 meter per seconds. These system also has a high degree of spatial orientation.

**36. Which of the following statement regarding lateral branch of the spinal nerve dorsal roots is NOT TRUE**

**A.** After synapsing in gray matter it joins dorsal column
**B.** Helps in eliciting local cord reflex
**C.** Proceeds through dorsal column to brain
**D.** Is a site of origin of spinocerebellar tract

**Ans. 36. (C)**

After entering the spinal cord the large myelinated fibres separate into lateral and medial branch. The lateral branch enters dorsal horn of gray matter and synapses with the neurons of the anterior and intermediate horn of gray matter.

**37. Which of the following statement about dorsal column tract fibres is TRUE**

**A.** Fibres from the lower part of the body lie in the lateral part of the spinal cord
**B.** Fibres from the tail end of the body is represented laterally in ventrobasal complex of thalamus

**C.** Fibres from the left side of the body represented in the left side of the thalamus

**D.** Fibres terminate in the temporal lobe

**Ans. 37. (B)**

Dorsal column tract has highest degree of spatial orientation. As the fibres are uncrossed at the spinal cord, the fibres from the lower half lies medially while that from upper part is present laterally. The fibres crosses in the medulla and hence left side is represented on the right side of the thalamus.

**38. Somatic sensory cortex exhibits highest degree representation for**

**A.** Lips **B.** Head

**C.** Thumb **D.** Trunk

**Ans. 38. (A)**

Sensory signal from all the modalities usually terminate in the cerebral cortex posterior to the central fissure. There are two sensory areas which lies in anterior parietal lobe. These are called somatic sensory area I and II. Because somatic sensory area I is very extensive it is popularly known as somatic sensory cortex. It has high degree of spatial orientation. The degree of representation depends on the number of the receptors in that area. The number of receptors in lips and are greatest, followed by face and thumb.

**39. The incoming signals enters the cortex through**

**A.** Neuronal layer 6

**B.** Neuronal layer 1

**C.** Neuronal layer 4

**D.** Neuronal layer 3

**Ans. 39. (C)**

Extending from the surface to the deeper layer cerebral cortex has six layers. Layer 1 is present on the surface and layer 6 is situated deep.

**40. Neuronal layer 4 of cerebral cortex**

**A.** Receives incoming sensory signals and spreads it to surface and deeper layers

**B.** Controls signals to basal ganglia, brainstem

**C.** Controls level of the excitation of the stimulated region

D. Coordinates between different areas of same as well as opposite side

**Ans. 40. (A)**

**41. Neuronal layer 1 and 2**

A. Receives incoming sensory signals and spreads it to surface and deeper layers

B. Controls signals to basal ganglia, brainstem

C. Controls level of the excitation of the stimulated region

D. Coordinates between different areas of same as well as opposite side

**Ans. 41. (C)**

These layers can facilitate the a specific region of the cortex through the diffuse and nonspecific input it receives from the lower brain centers.

**42. Neuronal layers 2 and 3**

A. Receives incoming sensory signals and spreads it to surface and deeper layers

B. Controls signals to basal ganglia, brainstem

C. Controls level of the excitation of the stimulated region

D. Coordinates between different areas of same as well as opposite side

**Ans. 42. (D)**

Neurons from these layers makes coonections with the related area of the cerebral cortex of the same as well as of the opposite site (via corpus callosum).

**43. Neuronal layer 5 and 6**

A. Receives incoming sensory signals and spreads it to surface and deeper layers

B. Controls signals to basal ganglia, brainstem

C. Controls level of the excitation of the stimulated region

D. Coordinates between different areas of same as well as opposite side

**Ans. 43. (B)**

The larger axons of the neurons from the layer 5 are projected to the distant area, e.g. basal ganglia, spinal cord, brainstem while the neurons of the layer 6 provides a feedback signals to thalamus.

**44. Injury to the somatic sensory area 1 does not affect**
**A.** Tactile discrimination
**B.** Localization of temperature and pain
**C.** Fine touch
**D.** Quality or intensity of pain

**Ans. 44. (D)**

**45. Deciphering of the incoming signal is done by**
**A.** Motor area
**B.** Somatic sensory association area
**C.** Somatic sensory area
**D.** Brodmann's area 3, 1 and 2

**Ans. 45. (B)**

Brodmann areas 5 and 7 constitutes somatic sensory association area. It is situated in the parietal cortex behind somatic sensory area 1. It receives signals from somatic sensory area 1, visual cortex, ventrobasal nuclei and other part of thalamus. It combines information from all these sources and hence in understanding the meaning of incoming signals.

**46. An injury to somatic sensory area [Brodmann's area 5 and 7] results in**
**A.** Amorphosynthesis
**B.** Ataxia
**C.** Poor localization
**D.** Muscular degeneration

**Ans. 46. (A)**

Removal of somatic association areas result in person ignoring opposite side of his body and loss of recognition of complex objects and complex forms by feeling them on the opposite side. Localization is primarily the function of somatic sensory area 1.

**47. Which of the following sensation is carried by dorsal column tract**
**A.** Crude touch
**B.** Unconscious kinesthetic sensation
**C.** Vibratory sensation
**D.** Pressure

**Ans. 47. (C)**

**48. Vibrations of 30-800 cycles per second are detected by**
**A.** Free nerve ending

**B.** Pacinian corpuscles
**C.** Meissener's corpuscles
**D.** Merkel's disk

**Ans. 48. (B)**

**49. Low frequency vibrations up to 80 cycles per second are detected especially by**

**A.** Free nerve ending
**B.** Pacinian corpuscles
**C.** Meissener's corpuscles
**D.** Merkel's disk

**Ans. 49. (C)**

**50. The impulse from pacinian corpuscles for vibratory sensation is carried by**

**A.** A δ fibres
**B.** C fibres
**C.** A α fibres
**D.** B fibres

**Ans. 50. (A)**

**51. The tickle and itch sensation are detected by**

**A.** Merkel's disk
**B.** Pacinian corpuscles
**C.** Meissener's corpuscles
**D.** Free nerve ending

**Ans. 51. (D)**

**52. Tickle and itch sensation is transmitted by**

**A.** Unmyelinated type C fibres
**B.** A- alpha fibres
**C.** B fibres
**D.** A- delta fibres

**Ans. 52. (A)**

Tickle and itch sensation are detected by a rapidly adapting mechanoreceptive free nerve endings. These receptors are present mostly in the superficial layer of the skin. Pain sensation can inhibit the sensation of itch through a phenomenon of lateral inhibition in the cord.

**53. Which of the following statement about antero-lateral tract is NOT TRUE**

**A.** It crosses to the opposite at the spinal cord level
**B.** It is composed of small, myelinated fibres
**C.** It has high degree of spatial orientation of nerve fibres with respect to their origin
**D.** It can transmit many modalities of sensation

**Ans. 53. (C)**

The second order neurons for the anterolateral pathways starts from lamina I, IV, V, VI. These tract has a poor spatial localization (more so for pain sensation) and less accuracy of the gradation of the intensities of the stimuli.

**54. Weber–Fechner principle is applied for the detection of**

**A.** Ratio of stimulus strength
**B.** Type of stimulus
**C.** Vibration frequency
**D.** Origin of stimulus

**Ans. 54. (A)**

Weber and Fechner had proposed that the discrimination of gradation of stimulus strength are is dependent on the logarithm of the stimulus strength. Recently it has been observed that this principle is more applicable to the higher intensity stimuli than lower intensity stimuli.

**55. The most important receptors for the detection of joint angulation in midranges of motion are**

**A.** Pacinian corpuscles
**B.** Golgi tendon organ
**C.** Expanded tip endings
**D.** Muscle spindles

**Ans. 55. (D)**

When the joints are moved, there is a stretch on some muscle and some are loosened, which is detected by muscle spindle. At the extremes of the movement the information is transmitted by pacinian corpuscles, Ruffini's nerve endings and some receptors from the tendons.

**56. Pacinian corpuscles and muscle spindles are adapted for detecting**

**A.** Rapid rates of changes
**B.** Pain
**C.** Slower rates of changes
**D.** Vibration

**Ans. 56. (A)**

**57. The anterolateral fibres originate mainly from**

**A.** Laminae I, IV, V, VI in the dorsal horns
**B.** Laminae I, VIII, IX, X in the dorsal horns

**C.** Laminae I, II, III,IV in the dorsal horns
**D.** Laminae I, II, IX, X in the dorsal horns

**Ans. 57. (A)**

After originating in these lamina the fibres cross in the anterior commissure to the opposite side anterior and lateral white column and ascend up. The fibres from lower part of the body lie laterally while those from upper part are present medially.

**58. The impulse transmission in the dorsal column medial leminiscal system has**

**A.** Velocity lower than the anterolateral system
**B.** Accurate gradation of intensity
**C.** Poor ability to transmit rapid repetitive signals
**D.** Poor spatial orientation

**Ans. 58. (B)**

Most of the sensation carried by dorsal column tract shows accurate gradation of intensity, being recognized in the 100 gradations of strength.

**59. Brainstem, thalamus and other associated basal regions helps in discrimination of**

**A.** Fine touch **B.** Pain
**C.** Vibration **D.** Stereognosis

**Ans. 59. (B)**

Other sensation requires a high degree of localization and discrimination and ends in the sensory cortex. While most of the pain fibres end in brainstem and thalamus and from here the signals are transmitted to the other basal areas of the brain and the somatic sensory cortex.

**60. The corticofugal signals**

**A.** Are transmitted from lower relay stations to cerebral cortex
**B.** Are inhibitory
**C.** Increases the transmission in relay nuclei
**D.** Are insensitive to sensory input

**Ans. 60. (B)**

**61. As related to the sensation of pain which of the following is TRUE**

**A.** Fast pain is carried by A-alpha fibres
**B.** Fast pain is felt in the deeper tissues

C. Slow pain is associated with little tissue destruction
D. Slow pain is prolonged

**Ans. 61. (D)**

Fast pain is sharp and acute caused by mechanical or themal stimuli, while slow pain is dull aching and long lasting and is caused mostly due to chemical and to certain extent by mechanical and thermal stimuli. Fast pain is transmitted by A–delta fibres while slow pain is transmitted by type C unmyelinated fibres.

**62. The thermal pain is felt when the skin is heated above**

A. 20°C B. 37°C
C. 45°C D. 40°C

**Ans. 62. (C)**

At the temperature of 45°C and above the tissue starts getting damage and this temperature the person begins to perceive the pain instead of the hot sensation.

**63. The chemical mediator responsible for pain is**

A. Glutamate B. Bradykinin
C. $K^+$ ion D. Substance P

**Ans. 63. (B)**

Glutamate and substance P are secreted in the spinal cord by A-delta fibres and type C fibres respectively. Substance P and prostaglandins increase the sensitivity of the pain endings but do not excite them directly.

**64. The fast pain signals**

A. Are elicited by chemical stimuli
B. Are transmitted in C fibres
C. Terminate in lamina I of spinal cord
D. Runs in dorsal column of spinal cord

**Ans. 64. (C)**

Fast pain fibres are mainly stimulated by themal and mechanical stimulation. After entering the spinal cord it ends in the lamina I (laminal marginalis) of the dorsal horns. Then it crosses to the opposite side and runs in the anterolateral column. It is transmitted via A-delta fibres.

**65. A neurotransmitter released by A-delta fibres in the spinal cord is**

A. Glutamate B. Substance P
C. Acetylcholine D. Serotonin

**Ans. 65. (A)**

Glutamate is the excitatory neurotransmitter widely used in the CNS. Its action lasts only for a few milliseconds. Substance P and probably also the glutamate is released in the spinal cord by the fibres carrying slow type of a pain.

**66. Fibres of Paleothalamic pathway terminate in all of the following area EXCEPT**

**A.** Sensory area II
**B.** Periaqueductal area
**C.** Mesencephalic tectal area
**D.** Brainstem reticular area

**Ans. 66. (A)**

Most of the slow fibres terminate in periaqueductal gray area, mesencephalic tectal area of the mesencephalon deep to the superior and inferior colliculi and brainstem reticular area. These areas appears to be important in the appreciation of suffering types of pains.

**67. All of the following are the component of the endogenous analgesic system EXCEPT**

**A.** Periaqueductal area
**B.** Raphe magnus nuclei
**C.** Dorsal horn of spinal cord
**D.** Caudate nucleus

**Ans. 67. (D)**

**68. Presynaptic inhibition by enkephalin is brought about by blockage of**

**A.** $Na^+$ channels in membrane of nerve terminals
**B.** $K^+$ in nerve membrane
**C.** Reuptake of neurotransmitter at synaptic gutter
**D.** $Ca^{++}$ channels in membrane of nerve terminals

**Ans. 68. (D)**

The neurons from periventricular nuclei in the hypothalamus and periaqueductal gray area terminating on raphe magnus nuclei secret enkephalins at their endings. The neurons from raphe magnus nuclei terminating on the dorsal horn of the spinal cord secret serotonin. This serotonin causes local neurons of the cord to release enkephalin. The enkephalin then causes pre as well as postsynaptic inhibition of type A-delta and type C fibres where they synapse in the dorsal horn.

**69. The transmission of pain impulse is depressed by stimulation of**

**A.** A- alpha fibres
**B.** Unmyelinated C fibres
**C.** A- beta fibres
**D.** B fibres

**Ans. 69. (C)**

A-beta fibres carry the tactile sensation. When these fibres are stimulated it inhibits the pain by the mechanism of the lateral inhibition.

**70. Stimulation of cold receptors starts at**

**A.** Very cold region **B.** 10°C – 15°C
**C.** 30°C – 49°C **D.** 45°C and above

**Ans. 70. (B)**

Below this temperature the cold sensation is perceived as pain sensation. When the temperature is so cold that the skin freezes even the pain sensation by the fibres carrying cold sensation is lost. At the temperature of 10°C–15°C the cold receptors starts getting stimulated and reaches the peak by 24°C and adding out slightly when the temperature rises above 40°C.

**71. Stimulation of warm receptors starts at**

**A.** Very cold region **B.** 10°C – 15°C
**C.** 30°C – 33°C **D.** 45°C and above

**Ans. 71. (C)**

The warm receptors starts getting stimulated at the temperature of 30°C and above and the response decreases when the temperature rises to 49°C. A temperature of 45°C causes the sensation of pain.

**72. Stimulation of pain receptors because of cold starts at**

**A.** Very cold region **B.** 10°C – 15°C
**C.** 20°C – 25°C **D.** 15°C – 20°C

**Ans. 72. (A)**

Refer to Ans 70.

**73. Stimulation of pain receptors because of heat starts at**

**A.** 15°C – 25°C **B.** 10°C – 15°C
**C.** 30°C – 40°C **D.** 45°C and above

**Ans. 73. (D)**

Refer to Ans 71.

**74. Which of the following statement regarding thermal receptors is FALSE**

**A.** Response fades away after a while after the stimulation

**B.** There are more warm receptors than cold receptors

**C.** Cold receptors are more in lips than in finger tips

**D.** Cold sensation is transmitted by A-delta fibres

**Ans. 74. (B)**

Cold and warmth receptors are located immediately beneath the skin. The number of cold receptors in the different areas of the body are 3–10 times more than the warm receptors. The number in the different areas vary from 15–25 cold points per square centimeter in lips to 3–5 cold points per square centimeter in the fingers to 1 cold point per square centimeter in trunk. The response to the change in the temperature is faster than the steady temperature. Though the adaptation is not 100%.

**75. Ventral lateral nuclei of thalamus controls the**

**A.** Circadian rhythm of the body

**B.** Behavioral functions of the body

**C.** Rapid directional movement

**D.** Reflex movements of the eyes and pupil

**Ans. 75. (B)**

**76. All of the following are characteristics of the conditioned reflex EXCEPT**

**A.** It is inborn

**B.** Is dependent on pre-existing unconditional reflexes

**C.** It can be unlearned

**D.** Hereditary does not play part in its transmission

**Ans. 76. (A)**

Conditioned is a learned response and requires a pre-existing unconditioned reflex. If the practice stops the response can be unlearned. Since it is learnt after the birth hereditary doesn't play a role in its development. Unconditioned reflex is inborn.

**77. Hippocampus helps in the**

**A.** Analysis of memory

**B.** Storage of new memory

C. Searching the memory stores
D. Memory recall

**Ans. 77. (B)**

**78. Wernicke's area helps in**
A. Analysis of memory
B. Storage of new memory
C. Searching the memory stores
D. Memory recall

**Ans. 78. (A)**

**79. Thalamus plays a role in**
A. Analysis of memory
B. Storage of new memory
C. Searching the memory stores
D. Memory recall

**Ans. 79. (C)**

**80. Mamillary body**
A. Helps in analysis of memory
B. Helps in storage of new memory
C. Is a site for projection of hippocampal fibres
D. Memory recall

**Ans. 80. (C)**

**81. Alzheimer's disease results from the loss of**
A. Dopaminergic fibres in the brain
B. Cholinergic fibres in the brain
C. Adrenergic fibres in the brain
D. Serotonergic fibres in the brain

**Ans. 81. (B)**

**82. Complete the connections of Papez circuit**

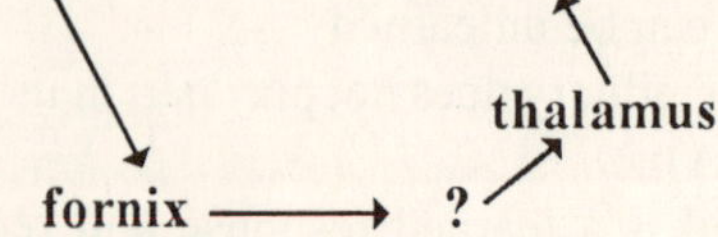


A. Mamillary body
B. Cerebellum
C. Basal ganglia
D. Corpus callosum

**Ans. 82. (A)**

The Papez circuit begins and ends in the hippocampus. Axons of the hippocampal pyramidal cells converge to form fimbria and finally the fornix. The fornix projects mainly to the mamillary body in the hypothalamus.The mamillary bodies in turn project to the anterior nuclei of thalamus by the way of mammilothalamic tract.The anterior nuclei project to the cingulated gyrus through the anterior limb of internal capsule. Then again through the cingulum and entorrhinal cortex the cingulated gyrus communicates with hippocampus.

**83. Klüver-Bucy syndrome is associated with all of the following EXCEPT**

**A.** Psychic blindness
**B.** Oral tendencies
**C.** Hypermetamorphosis
**D.** Excessive anger and fear

**Ans. 83. (D)**

Klüver-Busy syndrome results due to the bilateral ablation of amygdala. The animal is curious with the habit of putting everything in mouth, forgetfulness, loss of fear and decreased aggressiveness.

**84. Cerebellum helps in the controlling the voluntary movement by**

**A.** Coordinating the movements
**B.** Planning and sequencing the events
**C.** None of the above
**D.** Both A and B are correct

**Ans. 84. (D)**

For a person to carry-out it is important to plan and sequence the next few moves ahead (lateral zone). For carrying out the sequential movement the movements have to be coordinated (intermediate zone).

**85. Most of the damping function for the motor control system of the CNS is provided by**

**A.** Basal ganglia **B.** Red nucleus
**C.** Cerebellum **D.** Reticular formation

**Ans. 85. (C)**

Most of the movements of the body are pendular. To stop the movement the momentum has to be overcome. Cerebellum with appropriate learned subconscious signals prevents the overshot.

**86. In the absence of cerebellum, the ballistics movements**

A. Occur quickly
B. Doesn't have the initial surge
C. Force development of the movement is unaffected
D. Turn off of the movement is sudden

**Ans. 86. (B)**

Ballistics movements are the ones which are rapid and planned. Cerebellum helps in the initiation of the ballistics movement and providing the extra surge to the agonist muscles, development of the force for the movement and stopping the movement at the precise point.

**87. All of the following statement about the lateral zone of the cerebellum are true EXCEPT**

A. It doesn't receive direct input from periphery
B. It communicates with motor cortex
C. It communicates with primary and association somatic sensory area
D. It communicates with brainstem

**Ans. 87. (D)**

The lateral portion of the cerebellum doesn't have topographical representation of the body. It receives most of its input from the cerebral cortex. Through its connection with the association areas it helps in planning and sequencing the rapid movements.

**88. For the planning of the sequential movements, there is a bilateral communication between the lateral zone of cerebellum and**

A. Spinal cord
B. Premotor and sensory area of the cerebral cortex
C. Brainstem
D. Thalamus

**Ans. 88. (B)**

Refer to Answer 87.

**89. Lateral zone of the cerebellum is associated with all of the following EXCEPT**

A. Idea about the next sequential movement
B. Smooth progression of the movement
C. Rapid progression of movement
D. Posture

**Ans. 89. (D)**

Refer to answer 87.

**90. Overcompensation by the conscious part of the brain in the absence of cerebellum is called**

**A.** Dysmetria **B.** Nystagmus
**C.** Past pointing **D.** Dysarthria

**Ans. 90. (A)**

Dysmetria is the overshooting of the intended movements by the subconscious brain and overcompensation of the movement in the opposite direction by conscious brain. This is the effect seen because of the loss of the cerebellar control. This dysmetria (effect) leads to incoordination (result) of the movements. The lesions in the spinocerebellar can also cause dysmetria and ataxia.

**91. Incoordination of the movement resulting from damage to the cerebellum is called**

**A.** Ataxia
**B.** Past pointing
**C.** Dysdiadochokinesia
**D.** Rebound phenomenon

**Ans. 91. (A)**

Refer to Answer 90.

**92. Past pointing is a manifestation of**

**A.** Ataxia
**B.** Hypotonia
**C.** Dysmetria
**D.** Rebound phenomenon

**Ans. 92. (C)**

Past pointing means the movement exceeds the intended mark. This results from the failure to turn off the signal at the appropriate time.

**93. Cerebellar lesions results in**

**A.** Dysarthria **B.** Sensory aphasia
**C.** Dysphasia **D.** Motor aphasia

**Ans. 93. (A)**

Dysarthria is incoordination of speech. During the speech many muscle participates. It is important for these muscles to follow the orderly pattern. This orderly pattern is lost but the ability of the person to understand the speech and content to be delivered is not affected.

**94. Nystagmus with loss of equilibrium is commonly associated with the lesion of**

A. Vermis of the cerebellum
B. Intermediate zone of the cerebellum
C. Lateral zone of the cerebellum
D. Flocculonodular lobe of the cerebellum

**Ans. 94. (D)**

Flocculonodular lobe is the oldest part of the cerebellum. This lobe along with the vestibular system helps to maintain the equilibrium and balance of the body. Nystagmus is the tremors of the eyeballs that results from the failure of the damping function of the cerebellum.

**95. Rebound phenomenon in the cerebellar lesion occurs because of**

A. Exaggerated inverse stretch reflex
B. Loss of cerebellar component of the stretch reflex
C. Loss of the reciprocal inhibition
D. None of the above

**Ans. 95. (B)**

Rebound phenomenon results from the inability to brake the movement, which results in the exaggerated opposite movement. This is because of the stretch on the particular muscle stimulate the stretch receptors, which than causes the contraction of that muscle. In the absence of cerebellum when the stretch is removed the simultaneous correction cannot occur rapidly in the body resulting rebound phenomenon.

**96. All of the following nuclei are present in the cerebellum EXCEPT**

A. Dentate nuclei
B. Interpositus nuclei
C. Fastigial nuclei
D. Vestibular nuclei

**Ans. 96. (D)**

Though the vestibular nuclei are present in the medulla they functions as if they are the part of the cerebellum, because of their connection with flocculonodular lobe.

**97. A pathway (impulse, signal) arising from vermis of the cerebellum**

A. Passes through the fastigial nuclei
B. Goes to ventroanterior and ventrolateral nuclei of thalamus

**C.** Goes to dentate nuclei

**D.** Ends in cerebral cortex

**Ans. 97. (A)**

Vermis is the central portion of the cerebellum and is separated from the remaining portion of the cerebellum by the shallow grooves. The impulses from this region passes through fastigial nuclei to the medulla and pons. The neuronal circuit from this region functions in close association with vestibular nuclei and reticular formation of the brainstem to maintain equilibrium and posture respectively.

**98. All of the following statement about the pathway (impulse, signal) arising from intermediate zone of the cerebellum are true EXCEPT**

**A.** It passes through interpositus nuclei

**B.** It goes to red nucleus and reticular formation of the upper brainstem

**C.** It causes reciprocal contraction of agonists and antagonists

**D.** It helps in maintenance of equilibrium

**Ans. 98. (D)**

Intermediate zone is the part of the cerebellar hemisphere. The impulses arising from this region passes through the nucleus interpositus 1. to cerebral cortex through ventrolateral and ventroanterior nucleus of thalamus 2. to basal ganglia through midline structures of the thalamus 3. to red nucleus and reticular formation of upper brainstem. This circuit helps in coordinating the reciprocal contractions in agonists and antagonists of the distal part of the body (fingers, thumbs, feet, toes).

**99. All of the following statement about the pathway (impulse,signal) arising from the lateral zone of the cerebellum are true EXCEPT**

**A.** It passes through dentate nuclei

**B.** It ends in cerebral cortex

**C.** It helps in the coordination of the sequential motor activities

**D.** Helps in maintaining postural attitudes

**Ans. 99. (D)**

This region is also a part of the cerebellar hemisphere. The impulses from here pass through dentate nucleus and then through ventrolateral and ventroanterior nucleus of thalamus to cerebral cortex.

**100. The functional unit of cerebellum**

**A.** Is under the inhibitory control from brainstem
**B.** Is under the inhibitory control from periphery
**C.** Has its own output from deep cerebellar nuclei
**D.** Is under the excitatory control from Purkinje's cell

**Ans. 100. (C)**

Human cerebellar cortex is a large folded sheet and each fold is called a folium. Beneath this folded mass the cerebellar nuclei are present. Cerebellum has about 30 million nearly identical functional units. The functional unite centers on Purkinje's cells (30 million, cerebelllar cortex) and deep cerebellar nuclei.

**101. All of the following statement about the climbing fibres of the cerebellum are true EXCEPT**

**A.** It originates [receives input] from the inferior olivary nucleus
**B.** There is 1 climbing fibre for 10 Purkinje's fibres
**C.** Single action potential in climbing fibre will cause prolonged action potential
**D.** It ends in the granular layer

**Ans. 101. (D)**

After sending branches to the deep cerebellar nuclei the climbing fibre ends in molecular layer by synapsing with the Purkinje's cells (about 300 synapses). The prolonged action potential in the Purkinje's cell as a result of a single impulse in climbing fibre is called complex spike.

**102. All of the following statement about the mossy's fibres of the cerebellum are true EXCEPT**

**A.** It receives afferent from multiple sources
**B.** Synaptic connections are strong, producing prolonged action potential
**C.** It excites deep cerebellar nuclei
**D.** It ends in granular

**Ans. 102. (B)**

It receives input from higher brain, brainstem and spinal cord. After sending collaterals to the cerebellar nuclei the fibres form synapse with the granule cells in the granular layer. The axons of the granules cells extends to molecular layer and forms weak synapses with Purkinje's cells. Because of the weak synapse many mossy's fibres should be stimulated to cause an action potential in Purkinje's cell, but this action potential is of short duration and weak and is called simple spike.

**103. Which of the following statement about cerebellar firing at rest is TRUE**

**A.** Purkinje's cell and deep cerebellar nuclei fire continuously

**B.** Deep cerebellar nuclei fire about 50-100 APs per second

**C.** Purkinje's cell fire APs at a rate faster than deep cerebellar nuclei

**D.** The output from cerebellar nuclei is in favor of inhibition

**Ans. 103. (A)**

Deep cerebellar nuclei and Purkinje's cells fire continuously. Purkinje's cells fire 50–100 APs second while cerebellar nuclei fire at a still higher rates. Deep cerebellor nuclei are under excitatory influence of mossy's and climbing fibres and inhibitory influence of Purkinje's cells. At rest the balance is such that there is moderate level of excitation of the deep cerebellar nuclei.

**104. Which of the following statement about activation of cerebellum is TRUE**

**A.** Stimulation of deep cerebellar nuclei by climbing and mossy's fibres inhibits cerebellar activity

**B.** Purkinje's cell increases the activation of cerebellum

**C.** There is more excitation than inhibition under usual conditions

**D.** None of the above

**Ans. 104. (C)**

Refer to answer 103.

**105. Which of the following cerebellar cell is excitatory**

**A.** Granule cell **B.** Golgi cells

**C.** Basket cell **D.** Stellate cell

**Ans. 105. (A)**

Basket cells and stellate cells are present in the molecular layer. They are excited by the parallel fibres of the granule cells. These cells in turn cause the lateral inhibition of the Purkinje's fibres thus helping in sharpening the signals. These cells are also stimulated by the parallel fibres and thet in turn inhibit the granule cells. They help in limiting the duration of signal transmission in the cerebellar cortex.

**106. Destruction of the intermediate zone of the cerebellum is associated with**

- **A.** Hypotonia
- **B.** Intentional tremors
- **C.** Tremors at rest
- **D.** Dysdiadochokinesia

**Ans. 106. (B)**

Intermediate zone along with nucleus interpositus helps in the coordination of various movements.

**107. All of the following are the functions of the cerebellum EXCEPT**

- **A.** Rapid progression from one movement to another
- **B.** Timing of motor activities
- **C.** Instantaneous interplay between agonists and antagonists
- **D.** Sequencing of multiple, successive and parallel movements

**Ans. 107. (D)**

Basal ganglia is concerned with planning and controlling the complex patterns of movements, its intensity and its direction.

**108. All of the following are the functions of the basal ganglia EXCEPT**

- **A.** Controlling complex patterns of movements
- **B.** Controlling the relative intensities of the movements
- **C.** Controlling the direction of the movements
- **D.** Timing of motor activity

**Ans. 108. (D)**

Timing the motor activities is the function of the cerebellum.

**109. Cerebellum helps in motor activities by all of the following EXCEPT**

- **A.** By sequencing of motor activities
- **B.** By monitoring the motor activities

**C.** By making corrective adjustment
**D.** By its ability to cause muscle contraction

**Ans. 109. (D)**

Cerebellum does not have direct connections with the anterior horn cells. It affects the muscular tone via different descending tracts, e.g. vestibulospinal tract.

**110. Cerebellum helps in correction motor activities by all of the following EXCEPT**

**A.** Receiving updated input from motor cortex on the desired programme
**B.** Receiving continuous sensory input from peripheral part
**C.** Comparing the desired and the actual movement
**D.** Increasing the level of alertness

**Ans. 110. (D)**

Cerebellum receives the input from the periphery apprising cerebellum about the movement occurring and also from the motor cortex detail in cerebellum about the movements to be carried out. Cerebellum compares this two inputs and makes necessary correction for the smooth progression of the movement. Increasing the alertness is he function of reticular activating system.

**111. Flocculonodular lobe of cerebellum is concerned with**

**A.** Planning the sequential events
**B.** Monitoring motor activities
**C.** Timing the motor activities
**D.** Maintenance of equilibrium

**Ans. 111. (D)**

Flocculonodular lobe is connected to the equilibrium apparatus and vestibular system.

**112. Vermis of cerebellum is concerned with the movements of all of the following part EXCEPT**

**A.** Axial part of the body
**B.** Hip
**C.** Shoulder
**D.** Upper limb of the body

**Ans. 112. (D)**

**113. Intermediate zone of the cerebellum controls the movements of**

**A.** Axial part of the body
**B.** Hip

C. Shoulder
D. Upper limb of the body

**Ans. 113. (D)**

**114. Lateral portion of cerebellum**

A. Represents axial part of the body
B. Represents distal part of the body
C. Concerned with alertness
D. Connected with corresponding association areas of cerebral cortex

**Ans. 114. (D)**

This area of the cerebellum lacks the topographical representation of the body. By its connection with association areas it helps in the planning and sequencing the movements.

**115. An important pathway from the cerebral cortex to the cerebellum is**

A. Corticopontocerebellar tract
B. Olivocerebellar tract
C. Vestibulocerebello tract
D. Reticulocerebello tract

**Ans. 115. (A)**

This pathway terminates on the opposite lateral zone of the cerebellar hemisphere.

**116. Which of the following statement about the corticopontocerebellar tract is FALSE**

A. Originates from the sensory cortex
B. Ends in the opposite lateral zone of the cerebellar hemisphere
C. Originates from basal ganglia
D. Originates mainly from motor and premotor cortex

**Ans. 116. (C)**

**117. All of the following statement about dorsal spinocerebellar tract is true EXCEPT**

A. It enters cerebellum through inferior cerebellar peduncle
B. It terminates in vermis and intermediate zone of the cerebellum
C. It terminates in the ipsilateral side
D. It enters through superior cerebellar peduncle

**Ans. 117. (D)**

**118. All of the following statement about ventral spinocerebellar tract is true EXCEPT**

**A.** It enters cerebellum through inferior cerebellar peduncle

**B.** It terminates in vermis and intermediate zone of the cerebellum

**C.** It terminates in the ipsilateral side

**D.** Terminates bilaterally

**Ans. 118. (D)**

**119. The signal from dorsal spinocerebellar tract apprises cerebellum of all of the following EXCEPT**

**A.** Momentary status of the muscle contraction

**B.** Degree of tension on tendons

**C.** Position and the rate of movement

**D.** Motor signal at the anterior horn of the spinal cord

**Ans. 119. (D)**

Dorsal spinocerebellar tract receives it input mainly from muscle spindles and to a lesser extent from Golgi tendon organs, large tactile receptors of skin, joint receptors. Thus this tract apprises cerebellum of the muscle contraction (muscle spindles), tension in the tendon (Golgi tendon receptors) and rate of movement at the joint (joint receptor), force acting on surface of the body (tactile receptors).

**120. Which of the following is known as the silent are of the brain**

**A.** Somatic sensory cortex **B.** Basal ganglia

**C.** Cerebellum **D.** Broca's area

**Ans. 120. (C)**

It is known as silent area of the brain because the electrical stimulation cerebellum neither cause any sensation not any motor movement.

**121. Putamen circuit of basal ganglia is concerned with**

**A.** Cognitive functions

**B.** Execution of the learned pattern of movements

**C.** Sensory part of the speech

**D.** None of the above

**Ans. 121. (B)**

The input to the putamen comes mostly from premotor and supplementary motor cortex and the output goes mainly to the primary motor cortex.

**122. The neuronal circuit through putamen**

**A.** Begins mostly in primary motor cortex
**B.** Bypasses the caudate nucleus
**C.** Ends in the primary sensory area
**D.** Output goes mainly to supplementary motor cortex

**Ans. 122. (B)**

The input to the putamen comes mostly from premotor and supplementary motor cortex and to certain extent from somatic sensory cortex. The impulse from here than goes to the inner part of the globus pallidus but bypasses the caudate nucleus. From here the impulses are carried to thalamus and then to the primary motor cortex.

**123. Lesion in the globus pallidus leads to**

**A.** Athetosis
**B.** Hemibalismus
**C.** Chorea
**D.** Parkinson's disease

**Ans. 123. (A)**

There are three ancillary circuits working in close association with the primary putamen circuit. 1. putamen to external globus pallidus to subthalamus to thalamus to motor cortex 2. putamen to internal globus pallidus to substantia nigra to thalamus to motor cortex 3. external globus pallidus to subthalamus and back to external globus pallidus. Any lesion in the globus pallidus affects this circuits and leads to spontaneous, continuous writhing movements of a hand, arm, neck or face. This abnormal movement is called athetosis.

**124. All of the following statement about caudate nucleus circuit are true EXCEPT**

**A.** Neural connections extends into all lobes of the cerebrum
**B.** Most of the inputs are from association areas

**C.** Impulses ends in primary motor cortex
**D.** It is associated with cognition

**Ans. 124. (C)**

Caudate nucleus lies in the frontal lobe (anteriorly), occipital and parietal lobe (posteriorly) and temporal lobe. This helps caudate nucleus in receiving the inte-grated information from the association area thus helps in the thought (cognitive) process. The output is mostly to the prefrontal, premotor and supplementary motor cortex.

**125. Which of the following neurotransmitter is released in to nigrostriate fibres**

**A.** Dopamine **B.** Acetylcholine
**C.** Enkephalins **D.** Serotonin

**Ans. 125. (A)**

Dopamine is released by the fibres extending from substantia nigra to caudate nucleus and putamen. Putamen and caudate nucleus together is termed as striatum. Acetylcholine is released in to the fibres extending from cortex to striatum.

**126. Which of the following neurotransmitter always has inhibitory effect**

**A.** Acetylcholine
**B.** 5- hydroxytryptamine
**C.** Gamma-aminobutyric acid
**D.** Norepinephrine

**Ans. 126. (C)**

GABA is releases at the pathway from caudate nucleus and putamen to globus pallidus and substantia nigra. The GABA secreting neurons forms a negative feedback loop from cortex to basal gangia and back to the cortex. This helps in lending stability to the motor control system.

**127. Parkinson's disease is characterized by**

**A.** Dysdiadochokinesia
**B.** Hypotonia
**C.** Intentional tremors
**D.** Akinesia

**Ans. 127. (D)**

Akinesia is difficulty in initiating movement. It results from 1. decreased psychic drive to carry-out the movement due to dopamine secretion in the limbic system (nucleus accumbens). 2. Loss of dopaminergic neurons decreases the loss of inhibitory effect preventing the initiation and progression of the movement. Akinesia is also the most distressing effect seen in Parkinson's disease (paralysis agitans).

**128. Huntington's chorea is caused due to destruction of cell bodies**

**A.** Of dopamine secreting neurons in caudate nucleus and putamen

**B.** Of GABA secreting neurons in caudate nucleus and putamen

**C.** Of neurons in substantia nigra

**D.** Of neurons in globus pallidus

**Ans. 128. (B)**

Huntington's chorea is a hereditary disorder, caused due to destruction of cell bodies in of GABA secreting neurons in caudate nucleus and putamen and acetylcholinergic neurons in many parts of the brain. The GABA secreting neurons normally causes inhibition of globus pallidus and substantia nigra. The release of globus pallidus and substantia nigra from the inhibitory effect results in the abnormal movements. Huntington's chorea presents initially with flicking movements of the individual joints which later progresses to the entire body. This abnormal movement is also associated with dementia (due to loss of acetylcholinergic (Ach-ergic) neurons), motor dysfunctions.

**129. Dementia in Huntington's chorea results from the destruction of**

**A.** Acetylcholine secreting neurons

**B.** Serotonin secreting neurons

**C.** Norepinephrine secreting neurons

**D.** Enkephalin secreting neurons

**Ans. 129. (A)**

Dementia is due to loss of Acetylcholinergic neurons in the area of the cerebral cortex involved in thinking process.

**130. The term prosopagnosia (prosophenosia) means**

**A.** Inability to recognize face

**B.** Loss of taste sensation

**C.** Inability to talk
**D.** Loss of stereognosis

**Ans. 130. (A)**

Storage and recognition of faces is represented strongly in the right inferior temporal lobe in the right – handed people, though left lobe is also active. This part of the brain receives important input from visual signals. The lesion in this area results in inability to recognize the face but the patient can identify the person by his voice and also has the ability to recognize forms and reproduce them.

**131. The area of the brain responsible for higher intellectual functions is**
**A.** Motor speech area
**B.** Primary motor area
**C.** Primary sensory area
**D.** Sensory speech area (Wernicke's area)

**Ans. 131. (D)**

**132. Damage to the Wernicke's area results in inability to**
**A.** Recognize words
**B.** Hear the words
**C.** Comprehend the words
**D.** Read the words

**Ans. 132. (C)**

Wernicke's area is located at posterior end of the superior temporal gyrus. This area is also known as Gnostic area, general interpretative area, tertiary association area. It is highly developed in the dominant side of the brain. It is concerned with comprehension of auditory and visual information. Lesion in this area results in fluent aphasia (abnormalities of language functions which are not due to defects of auditory or visual or motor system. In this type of aphasia patient fails to understand the meaning of spoken or written words. The speech (ability to talk) is normal but it lacks meaning and is full of neologism and jargon.

**133. All of the following statement about the sensory interpretative areas are true EXCEPT**
**A.** It is located on in the posterior part of the superior temporal lobe
**B.** It is highly developed in the non-dominant side of the brain

C. It plays important role in intelligence
D. It is the meeting point of somatic, visual, auditory areas

**Ans. 133. (B)**

**134. Dyslexia**

A. Is a loss of ability to see the words
B. Results from damage to Broca's speech area
C. Is a loss of ability to interpret the meaning of the written words
D. Is a loss of ability to hear words

**Ans. 134. (C)**

According to some. The term dyslexia means word blindness, i.e. inability to interpret the written words though they can see the words while according to some dyslexia is the broad term applied to impaired ability to read (Ganong). The cause appears to be the damage to angular gyrus (most anterioinferior portion of the posterior parietal and fusing posteriorly with the visual arcas of occipital lobe) lobe behind Wernicke's area. This disorder is due to an inherited abnormality affecting about 5% of the population. The pathology involves either the reduced ability to recall the speech sounds leading to difficulty in translating the them in to the speech units (phonemes) or slowing of translation process due to defect in the magnocellular part of the visual system.

**135. All of the following statement about the slow wave sleep are true EXCEPT**

A. It is associated with the slow waves in EEG
B. It is associated with deep and restful sleep
C. It is associated with active dreaming
D. It is easy to wake up person in this sleep than in REM sleep

**Ans. 135. (D)**

The slow wave sleep is so called because the brain waves during this sleep are very slow. Though this sleep is associated with dreaming the dreams are usually not remembered because the memory does not get consolidated.

**136. All of the following statement about the REM sleep are true EXCEPT**

A. It occupies 25% of the total sleep time
B. It occurs every 90-120 minutes

**C.** It is associated with increased metabolism of brain

**D.** It is associated with increased muscle tone

**Ans. 136. (D)**

Rapid eye movement sleep is also known as paradoxical sleep. Though the person is asleep there is increased activity of the brain but this activity is not channelised. REM sleep occupies about 75-80% and 45-50% of the total sleep duration in premature and in full-term babies respectively. In adult it occupies about 25% of the total sleep and than in the old age the duration falls further. Though there is rapid eye movement the tone of the muscle especially voluntary muscle (locus ceruleus dependent) decreases.

**137. Stimulation of which of the following would induce sleep**

**A.** Preoptic area

**B.** Superior colliculus

**C.** Inferior colliculus

**D.** Nucleus tractus solitarius

**Ans. 137. (A)**

There appears to be three areas, the stimulation of which would induce sleep. 1. diencephalic sleep zone (posterior hypothalamus and some thalamic nuclei, low frequency stimulation 2. medullary synchronizing zone (reticular formation of medulla at the level of nucleus tractus solitarius, low frequency) 3. basal forebrain sleep zone (preoptic area and diagonal band of Broca, high as well as low frequency).

**138. Which of the following would be present in the CSF and blood of the animal kept awake for longer periods**

**A.** Prostaglandin $E_2$ **B.** Muramyl peptide

**C.** Acetylcholine **D.** Serotonin

**Ans. 138. (B)**

The serotonin agonists suppress sleep. Many substance are claimed to be promoting sleep. Adenosine, $PGD_2$ and a putative sleep peptide (muramyl peptide).

**139. Damage to which of the following would result in decreased frequency and duration of REM sleep**

A. Intralaminar nuclei of thalamus
B. Reticular activating system
C. Substantia nigra
D. Locus ceruleus

**Ans. 139. (D)**

There is increase in the activity in pontine area, in amygdala and in anterior cingulate gyrus during the REM sleep and there is decrease in the activity in prefrontal and parietal cortex. But the recent research indicates that noradrenergic neurons in locus ceruleus causes wakefulness and that these neurons are silent during REM sleep.

**140. The alpha waves of electroencephalogram (EEG)**

A. Occurs at a frequency of 80-130 Hz
B. Are seen during active brain
C. Are recorded mainly from occipital lobe
D. Are recorded from frontal lobe

**Ans. 140. (C)**

Alpha waves are most marked in the parieto-occipital region. It is observed during quiet and restful period with closed eyes and wandering mind. It occurs at a frequency of 8–12 Hz and has an amplitude of about 50–100 microvolts when recorded from scalp.

**141. All of the following statement about the beta waves of EEG are true EXCEPT**

A. These waves occur at a frequency of 14-25 Hz
B. These waves are recorded from parietal and frontal lobe
C. These waves are seen during rest
D. These waves are seen during increased mental stress

**Ans. 141. (C)**

The beta waves replace alpha waves moment there is extra activation of brain or during tension or when the eyes are opened.

**142. All of the following statement about sleep are true EXCEPT**

A. 2nd stage of the sleep is characterized by sleep spindles
B. REM sleep is associated with beta waves
C. Deep sleep is also called desynchronized sleep
D. Delta waves are seen during the 4th stage of deep sleep

**Ans. 142. (C)**

The stage I of deep sleep is characterized by low amplitude, high frequency waves in EEG, stage II is characterized by the waves with the frequency of 10–14 Hz and an amplitude of 50 microvolts (sleep spindles) (Ganong). The stage III results in waves with even lower frequency but increased amplitude. The stage IV waves occurs a frequency of 1–3 waves per second and has a high amplitude (delta waves). Thus the slow wave sleep shows the marked synchronization of the waves.

**143. The waves of EEG**

- **A.** results from non-synchronised nerve discharge
- **B.** Are generated from 3rd and 4th neuronal layer of the cerebral cortex
- **C.** Are the summated potential of different nerve fibres
- **D.** Depends on number of neurons firing impulse

**Ans. 143. (C)**

It results from the summation of the synchronised discharge of the different neurons. The non-synchronised discharge often nullify each other's electrical activity.

**144. Petit mal epilepsy is characterized by**

- **A.** Presence of spike and dome pattern in the EEG waves
- **B.** Presence of synchronized high voltage discharge
- **C.** Post seizure depression of the CNS
- **D.** Presence of tonic-clonic convulsions

**Ans. 144. (A)**

Petit mal epilepsy is characterized by a brief spell (3– 30 seconds) absenteinism (of unconsciousness or diminished consciousness with several twitch-like contraction of the muscles especially of the head region followed by return of consciousness and resumption of activities). It is commonest in the older children and usually disappears by the age of 30. It involves the basic thalamocortical brain activating system.

**145. Which of the following statement about grand mal epilepsy is TRUE**

- **A.** It is associated with low voltage, synchronized discharge

**B.** It is associated with twitch like muscular contraction
**C.** It is commonest in children
**D.** The abnormal neuronal circuit is present in the basal region of the brain

**Ans. 145. (D)**

Grand mal epilepsy results from the high voltage, synchronized discharge in the neuronal circuit of the basal brain. It spreads to the entire cortex, spinal cord. This results in tonic-clonic convulsions of the body.

**146. Psychomotor epilepsy (type of focal epilepsy) is characterized by all of the following EXCEPT**
**A.** Sudden anxiety and fear
**B.** An episode of quiescence
**C.** Loss of memory for a short period
**D.** Slurred speech

**Ans. 146. (B)**

It is associated with the period of abnormal rage.

**147. All of the following can precipitate the attack of grand mal epilepsy EXCEPT**
**A.** Intense emotional trauma
**B.** Acidosis
**C.** Strong auditory or visual stimuli
**D.** Increased body temperature

**Ans. 147. (B)**

Acidosis is known to decrease the excitability of the nerve and synapse.

**148. Depression of mood can result from diminished activity of**
**A.** Acetylcholinergic neurons
**B.** Noradrenergic neurons
**C.** Dopaminergic neurons
**D.** None of the above

**Ans. 148. (B)**

The axons of the noradrenegic neurons from the locus ceruleus extends to limbic cortex, thalamus and cerebral cortex. Excitation of this neurons results in mood elevation and the drugs (reserpine) which blocks this neurons are known to cause depression.

**149. Tricyclic antidepressant elevates the mood by increasing the excitatory effect of noradrenaline by**

**A.** Increasing the release of noradrenaline from presynaptic membrane
**B.** Decreasing the reuptake of noradrenaline by presynaptic membrane
**C.** Slowing the rate of destruction of noradrenaline by MAO
**D.** All of the above

**Ans. 149. (B)**

Rigidity is the increased tone of the muscles. Tone in the muscle is maintained because of the balance in the inhibitory and excitatory input to the anterior horn of the spinal cord. Lesion at the midbrain results in loss of the excitatory input to the inhibitory descending tacts which results in unopposed action of the excitatory descending tracts.

**150. Decerebrate rigidity is caused due to the lesion in**

**A.** Cerebellum
**B.** Medulla oblongata
**C.** Somatic sensory cortex
**D.** Midbrain

**Ans. 150. (D)**

Rigidity is increased tone on the muscle. Cerebellar lesion causes hypotoma. Lesion in mid brown distrupts the balance between inhibitory and excitatory impulses to muscle the excitatory impulse generation is in born but inhibitory impulses to muscle depends on activation by higher centres lesion in mid brain leads to distruption it impulses from higher centres to inhibitory impulses.

# 10
# Special Sensation

**1. Light travels through air medium at a velocity of**

A. 380 m/s
B. 300,000 km/s
C. 300,000 m/s
D. 380 km/s 380 m/s

**Ans. 1. (B)**

**2. A light travels faster through**

A. Air
B. Solid
C. Liquids
D. Glass

**Ans. 2. (A)**

**3. The refractive index of air is**

A. 1.33
B. 1.4
C. 1.0
D. 1.38

**Ans. 3. (C)**

The refractive index is the ratio of the speed of the light in air to that in the medium.

**4. The refractive index of cornea is**

A. 1.33
B. 1.4
C. 1.0
D. 1.38

**Ans. 4. (D)**

**5. The refractive index of aqueous humor is**

A. 1.33
B. 1.4
C. 1.0
D. 1.38

**Ans. 5. (A)**

**6. The refractive index of crystalline lens is**

A. 1.34
B. 1.4
C. 1.0
D. 1.38

**Ans. 6. (B)**

**7. The refractive index of vitreous humor is**

A. 1.34
B. 1.4
C. 1.0
D. 1.38

**Ans. 7. (D)**

**8. The total refractive power of the reduced eye is**

**A.** 69 diopters **B.** 80 diopters
**C.** 29 diopters **D.** 59 diopters

**Ans. 8. (D)**

Reduced eye is a simple and schematic representation of the optics of the eye by algebraric sum of the all the refractive surfaces of eye. This single media has a refractory index of 1.336. The refractive surface is considered to exist 17 mm (1.4 cm behind cornea) in front of the retina.

**9. All of the following changes occur in an eye accommodate for the near vision EXCEPT**

**A.** Lens becomes convex
**B.** Increased diopteric strength of the lens
**C.** Constriction of pupils
**D.** Convergence of the eye ball

**Ans. 9. (A)**

Accommodation is the changes occurring in the eye in order to focus the object properly. At rest the eyes are accommodated for the far vision. In order to see the near object clearly the anterior curvature of the eye increases which increases the refractive power of the eye, the pupil constricts to increase the depth of focus and both the eyes converge.

**10. A myopic person**

**A.** Is unable to see the near object clearly
**B.** Needs bifocal lens
**C.** Needs convex lenses
**D.** Is unable to see the far object clearly

**Ans. 10. (D)**

Myopia is the error of refraction in which person is unable to see the far things clearly but can see the near objects clearly. This error is common in youngster and is caused either by increased curvature of the lens or long eyeball. The image is formed in front of the retina and this can be corrected by using a concave lens.

**11. A total refractive power of the reduced eye when a lens is accommodate for the far vision is**

**A.** 59 diopters **B.** 50 diopters
**C.** 29 diopters **D.** 40 diopters

**Ans. 11. (A)**

Most of the refractory power is provided by the anterior curvature of cornea and not by the lens. This is because of the difference in the refractory indices of cornea and air.

**12. Most of the refractive power of the eye is provided by**

A. Retina
B. Anterior surface of cornea
C. Aqueous humor
D. Crystalline lens

**Ans. 12. (B)**

Refer to answer – 11. the lens provides refractory power of only 20 diopters as it is surrounded by the fluid whose refractory power is not very different from that of the lens.

**13. The total refractive power of crystalline lens of the eye is**

A. 5-10 diopters
B. 55-60 diopters
C. 16-20 diopters
D. 25-30 diopters

**Ans. 13. (C)**

Refer to answer – 12.

**14. Decreased accommodative power of the eyes due to advancing age is**

A. Short sightedness
B. Presbyopia
C. Astigmatism
D. Long sightedness

**Ans. 14. (B)**

With the advancing age especially after the age of 40 years the elastic tissue of the lens undergoes degeneration. This affects the diopteric power of the lens because of decreased ability to change its curvature. This affects the focusing of near objects.

**15. The term eumetropia means**

A. Difficulty in focusing far object
B. Difficulty in focusing near object
C. Failure of accommodation of eyes
D. Eyes with normal vision

**Ans. 15. (D)**

The eyes with a vision of 6/6 without the use of glasses is called an Eumetropic vision.

**16. The term myopia means**

**A.** Difficulty in focusing far object
**B.** Difficulty in focusing near object
**C.** Failure of accommodation of eyes
**D.** Eyes with normal vision

**Ans. 16. (A)**

**17. The term presbyopia means**

**A.** Difficulty in focusing far object
**B.** Difficulty in focusing near object
**C.** Failure of accommodation power of eyes
**D.** Eyes with normal vision

**Ans. 17. (C)**

Because of the loss of accommodative power the person has difficulty in seeing near objects clearly but the far objects can be seen clearly.

**18. The term hypermetropia means**

**A.** Difficulty in focusing far object
**B.** Difficulty in focusing near object
**C.** Failure of accommodation of eyes
**D.** Eyes with normal vision

**Ans. 18. (B)**

Hypermetropia (hyperopia, far sightedness) can result from either too short eyeball or a weaker lens system which results in the decreased diopteric power of the eye. Thus the image is formed behind the retina. This people have problems in sighting a near object but can see the far objects clearly. It commonly seen in old people.

**19. An ability of the visual apparatus to perceive the distance of an object is called**

**A.** Depth of focus **B.** Depth perception
**C.** Accommodation **D.** Visual acuity

**Ans. 19. (B)**

**20. The depth perception is achieved by all of the following EXCEPT**

**A.** Size of the image of known object on retina
**B.** Stereopsis
**C.** Accommodation
**D.** Phenomenon of moving parallax

**Ans. 20. (C)**

The prior knowledge of the size helps to locate the distance as the previous information is stored by brain. Stereopsis is the perception of the parallex by the binocular vision. Persons with binocular vision has a better ability of depth perception than person with a single eye vision. The images of the object close to our eyes move rapidly while images of the distant object hardly moves when the head is turned. This phenomenon is called the moving parallex.

**21. The rate of formation of aqueous humor in an eye is**

**A.** 2–3 microlitre/min　　**B.** 2–3 ml/min

**C.** 20–30 microlitre/min　　**D.** 20–30m/min

**Ans. 21. (A)**

Aqueous humor is a clear optically transparent fluid which fills the anterior and the posterior chamber. Its pH is 7.53, its osmotic pressure is higher than that of blood and the specific gravity is slightly more than the water. It provides the oxygen and nutrition to lens and cornea (both are avascular) and carries away the metabolic products.

**22. Most of the aqueous humor is secreted by**

**A.** Lens　　**B.** Choroids plexus

**C.** Iris　　**D.** Ciliary process

**Ans. 22. (D)**

**23. Total surface area of ciliary process in each eye is**

**A.** 10 $cm^2$　　**B.** 6 $cm^2$

**C.** 6 $mm^2$　　**D.** 20 $cm^2$

**Ans. 23. (B)**

The anterior part of the inner surface of the ciliary body is thrown in to many folds. These folds are called ciliary process which secretes aqueous humor. Because of the fold the ciliary process has a larger surface area. Beneath these processes is a vascular area and the surface is covered by a secretory epithelium.

**24. Aqueous humor is formed by the process of**

**A.** Active secretion

**B.** Osmosis

**C.** Passive diffusion
**D.** Facilitated diffusion

**Ans. 24. (A)**

Most (75%) of the aqueous humor is formed by the active secretion by the ciliary process epithelium. But ultrafiltration and pinocytosis also helps in the formation of aqueous humor.

**25. The canal of Schlemm is a**

**A.** Capillary **B.** Arteriole
**C.** Thin walled vein **D.** Artery

**Ans. 25. (C)**

**26. The normal intraocular pressure ranges from**

**A.** 30–40 mm Hg **B.** 12–20 mm Hg
**C.** 4–8 mm Hg **D.** 40–50 mm Hg

**Ans. 26. (B)**

Intraocular pressure (IOP) is the pressure of the aqueous humor in the anterior chamber. The various factors affecting IOP are age, hereditary, posture, hormones, exercise, length of the eyeball.

**27. Tonometry is a process of measuring**

**A.** CSF pressure
**B.** Blood pressure in capillaries
**C.** Pleural pressure
**D.** Intraocular pressure

**Ans. 27. (D)**

**28. The level of intraocular pressure is determined mostly by**

**A.** The resistance to the outflow of aqueous humor from posterior to anterior chamber
**B.** The resistance to the outflow of aqueous humor from anterior chamber to canal of Schlemm
**C.** Episcleral venous pressure
**D.** None of the above

**Ans. 28. (B)**

The resistanct to the flow through canal of Schlemm is principally because of the meshwork of trabeculae in the canal of Schlemm. These trabeculae have openings which have diameter of only 2–3 micrometers. To a smaller extent episcleral vessel pressure can also impede the drainage and affect the IOP.

**29. The light–sensitive photochemical is found in**

A. The nucleus of rods
B. Synaptic body of rods
C. Outer segment of rods
D. Inner segment of rods

**Ans. 29. (C)**

Rods and cons are the receptors for the vision. Both rods and cons are have outer segment, inner segment, nucleus and the synaptic body. Rods has photosensitive pigments called rhodopsin, which are responsible for the vision of the white light while cons has photosensitive pigments called color pigments which are responsible for color vision.

**30. Which of the following is an activated rhodopsin**

A. Metarhodopsin II
B. Metarhodopsin I
C. Lumirhodopsin
D. Bathorhodopsin

**Ans. 30. (A)**

When a light falls on retina cis-retinal form gets converted in to all-trans retinal form resulting in the formation of bathorhodopsin, which gets converted to lumirhodopsin, which gets decayed to metarhodopsin I which finally decays to form metarhodopsin II. This metarhodopsin II brings about electrical changes in the rods.

**31. When there is a decomposition of rhodopsin there is**

A. Decreased membrane conductance for $K^+$ in rods
B. Decreased membrane conductance for $Na^+$ in outer segment of rods
C. Increased membrane conductance for $Na^+$ in outer segment of rods
D. Decreased membrane conductance for $Na^+$ in inner segment of rods

**Ans. 31. (B)**

The inner segment pumps the sodium ions out actively. In the dark the sodium ions diffuses inside from the outer segment. This decreases the membrane potential to – 40 to –50 mv when light strikes the retina the sodium channels in outer segment closes thus sodium influx is stopped which makes the membrane potential more negative.

**32. Which of the following statement is NOT TRUE**

**A.** The photon activates an electron in 11 - cis retinal portion of the rhodopsin

**B.** The activated rhodopsin functions as an enzyme to activate transducin

**C.** Transducin activates phosphodiesterase

**D.** Activated phosphodiesterase causes formation of cGMP

**Ans. 32. (D)**

cGMP binds with the sodium channels in the outer segment keeps it open. Phosphodiesterase hydrolyses the cGMP and thus results in the closure of the sodium channels in the outer segment.

**33. During dark adaptation**

**A.** Rods adapt earlier and faster than cones

**B.** The pupillary size decreases

**C.** Cones achieves higher sensitivity than rods

**D.** Neural adaptation occurs very fast

**Ans. 33. (D)**

The eyes gets adapted to the light and dark by 1. change in the concentration of the photosensive pigments 2. change in the papillary size 3. change in the firing rate of neurons in the visual chain (neural adaptation). Though the adaptation caused by the neural adaptation contributes to a small degree it occurs in a fraction of a second. Cons are fast to adapt but the degree of sensitivity achieved by cons is lesser than that is achieved by rods. The papillary size increases during dark adaptation in order to allow passage for entry of more light.

**34. Protonamaly is a condition in which**

**A.** Red cones are absent

**B.** Green cones are absent

**C.** Blue colour is confused with yellow colour

**D.** Blue cones are absent

**Ans. 34. (A)**

The specificity of the cones to a particular colour depends on the type of colour pigment present in it. There are three basic colour pigments, blue sensitive pigments, green sensitive pigments, red sensitive pigments. Red sensitive pigments peak absorbancy at the wavelength of 570 nanometers. When a single group of colour sensitive cone is absent the person is unable to distinguish some colours from others. Red blindness is a genetic disorder mostly present in males.

**35. Deuteronamaly is a condition in which**

**A.** Red cones are absent
**B.** Green cones are absent
**C.** Blue colour is confused with yellow colour
**D.** Blue cones are absent

**Ans. 35. (B)**

Green sensitive pigments peak absorbancy at the wavelength of 535 nanometers.

**36. Tritanomaly is a condition in which**

**A.** Red cones are absent
**B.** Green cones are absent
**C.** Blue colour is confused with yellow colour
**D.** Blue cones are absent

**Ans. 36. (D)**

Blue sensitive pigments peak absorbancy at the wavelength of 445 nanometers.

**37. In Tetranope is**

**A.** Red cones are absent
**B.** Green cones are absent
**C.** Blue colour is confused with yellow colour
**D.** Blue cones are absent

**Ans. 37. (C)**

**38. The retrograde signal transmission in retina is done by**

**A.** Horizontal cells **B.** Amacrine cells
**C.** Interplexiform cells **D.** Ganglion cells

**Ans. 38. (C)**

These cells transmits signal from inner plexiform layer to the outer plexiform layer. These are mostly inhibitory and their probable role is to control the degree of contrast in the visual image.

**39. Which of the following statement regarding visual pathway is FALSE**

**A.** Retina has a dual pathway

B. Visual pathway by cone has large and fast nerve fibres
C. Number of neurons in rods and cones pathway are same
D. Visual pathway by rods is faster

**Ans. 39. (D)**

Retina has rod based vision (old type) and cone based vision (new vision). Visual signals from cones is transmitted 3–5 times faster than that from the rods.

**40. Only neurons in retina, transmitting the visual signals as action potential are**

A. Bipolar cells
B. Ganglion cells
C. Horizontal cells
D. Amacrine cells

**Ans. 40. (B)**

All other retinal neuronal cells conduct their signal by electrotonic conduction. This (electrotonic conduction) allows graded conduction of the signal strength. Sometimes the action potentials can be recorded from the amacrine cells.

**41. In retina**

A. Central part has greater sensitivity for weak light
B. Peripheral part has high degree of visual acuity
C. Central most part has only cones
D. Rods are more sensitive to colour light than cons

**Ans. 41. (C)**

Fovea centralis which has about 35,000 cones.

**42. The W ganglionic cells of retina**

A. Transmit signals at a velocity of 80 m/sec
B. Receives most of their excitation from cones
C. Has narrow field in retina
D. Detect directional movement

**Ans. 42. (D)**

The W ganglionic cells constitute about 40% of the total ganglionic cells (1.6 millions). These receives their excitation mainly from rods and transmit signals at a speed of 8 m/sec.

**43. Which of the following statement about the X ganglionic cells of retina is not TRUE**

**A.** These are the fastest conducting ganglionic cells cells
**B.** Detect colour vision
**C.** Has narrow field in retina
**D.** Are the most abundant cells

**Ans. 43. (A)**

These constitutes about 55% of the total ganglionic cells. These have a small field because their dendrites do not spread wide. These transmits impulse in the optic nerve at a rate of 14 m/sec.

**44. The Y ganglionic cells of retina**

**A.** Are smallest of all cell types
**B.** Apprise the CNS instantly about visual event
**C.** Are most abundant cell type
**D.** Responds to rapid changing image

**Ans. 44. (B)**

These are the largest and the fastest ganglionic cell with a diameter of about 35 micrometers and conduction velocity of more than 50 m/sec. This helps them to quickly apprise the CNS about the visual events. These constitute only about 5% of the total ganglionic cells.

**45. Suprachiasmatic nuclei of hypothalamus controls the**

**A.** Circadian rhythm of the body
**B.** Behavioral functions of the body
**C.** Rapid directional movement
**D.** Reflex movements of the eyes and pupil

**Ans. 45. (A)**

**46. Pretectal nuclei of brain controls the**

**A.** Circadian rhythm of the body
**B.** Behavioral functions of the body
**C.** Rapid directional movement
**D.** Reflex movements of the eyes and pupil

**Ans. 46. (D)**

**47. Superior colliculus of mid brain controls the**

**A.** Circadian rhythm of the body
**B.** Behavioral functions of the body
**C.** Rapid directional movement
**D.** Reflex movements of the eyes and pupil

**Ans. 47. (C)**

**48. The 3rd dimensional position of visual objects in the space around the body is transmitted mainly from**

**A.** A large Y optic nerve fibres of Y ganglionic cells
**B.** X optic nerve fibres
**C.** Fibres of W ganglion cells
**D.** None of the above

**Ans. 48. (A)**

**49. Lesion in the central part of optic chiasma leads to**

**A.** Anopia in both the eyes
**B.** Binasal hemianopia
**C.** Bitemporal heteronomous hemianopia
**D.** Loss of only macular vision

**Ans. 49. (C)**

**50. All of the following are the functions of the middle ear EXCEPT**

**A.** Impedance matching
**B.** Protection of damage to cochlea
**C.** Pitch discrimination
**D.** Filtration of sound frequencies

**Ans. 50. (C)**

Pitch discrimination is the function of the auditory cortex which is present in the temporal lobe (area 41).

**51. Which of the following statement about impedance matching is FALSE**

**A.** It is due to the difference in surface area of tympanic membrane and foot plate of stapes
**B.** It is due to the lever system of the ossicles of the middle ear
**C.** Is due to the contraction of the stapedius muscle
**D.** It is 50–75% perfect for the sound frequency between 300–3000 cycles per second

**Ans. 51. (C)**

Contraction of the stapedius helps in the acoustic reflex which prevents the damage to the cochlea from the loud sounds.

**52. Which of the following sentence is TRUE**

**A.** Basilar fibres at the base of the cochlea are taller than those at the apex of the cochlea

**B.** The basilar fibres at the apex are more stiffer than those at the base of the cochlea

**C.** Basilar fibres at the base vibrate best at high frequency

**D.** None of the above

**Ans. 52. (C)**

There is a progressive increase in the length of the basilar fibres from base to the apex while the diameter of the fibres decreases from base to the apex. So the fibres present at the base are short and stiff, while the fibres at the apex are long and limber which vibrates at the low frequency.

**53. The taste buds for the sweet and salty tastes are located**

**A.** Lateral sides of the tongue

**B.** Posterior tongue

**C.** Tip of the tongue

**D.** Soft palate

**Ans. 53. (C)**

# 11
# Endocrine

**1. The non-diffusible calcium constitutes**
**A.** 40-45% of the total serum calcium
**B.** 10-17% of the total serum calcium
**C.** 45-50% of the total serum calcium
**D.** 20-30% of the total serum calcium

**Ans. 1. (A)**

Total body calcium is 1100 mg. Out of this 99% is present in the skeleton and teeth. The normal serum calcium level is 9–11 mg% (2mg% of calcium = 1mEq/L, 4mg% of calcium = 1 mmol /L). Out of this 40–45% is bound to plasma proteins and cannot diffuse out of the vascular compartment.

**2. The bound and diffusible form of calcium constitutes**
**A.** 4-5% of the total serum calcium
**B.** 10-12% of the total serum calcium
**C.** 45-50% of the total serum calcium
**D.** 20-30% of the total serum calcium

**Ans. 2. (B)**

The 10-12% of the serum calcium is bound to ions like citrates and phosphates. In this form it can easily diffuse but does not play a role in normal physiological functions of the body.

**3. The ionized and diffusible calcium constitutes**
**A.** 30-40% of the total serum calcium
**B.** 10-17% of the total serum calcium
**C.** 45-50% of the total serum calcium
**D.** 80-90% of the total serum calcium

**Ans. 3. (C)**

It is this form which participates in the physiological functions of the body. Fall in the level of ionized calcium in serum results in abnormal effects in body.

**4. Osteoblasts are**

**A.** Mitotic in nature

**B.** Bone eating cells

**C.** Releases lysosomal enzymes

**D.** Synthesize and secrete collagen

**Ans. 4. (D)**

Osteoblasts are the bone forming cells. These are the modified fibroblasts.It secrets collagen - I which helps in formation of the bone matrix.

**5. Osteoclasts**

**A.** Are nonmitotic cells

**B.** Has receptors for parathyroid hormone

**C.** Has receptor for calcitonin hormone

**D.** Release osteocalcin

**Ans. 5. (C)**

Osteoclasts are the bone eating cells. These belongs to the monocytic family. Though it breaks down the bone matrix it does has the receptors for calcitonin but not for parathyroid hormones.

**6. Parathyroid hormone has all of the following effect on kidney EXCEPT**

**A.** Increased reabsorption of calcium

**B.** Increased reabsorption of phosphorus

**C.** Increased reabsorption of magnesium

**D.** Increased activity of 1 - alpha hydroxylase

**Ans. 6. (B)**

Parathyroid hormone increases the excretion of phosphates from by proximal convoluted tubules. The activity of the enzyme 1- alpha hydroxylase is increased by parathyroid hormones. This enzyme helps in conversion of 25-hydroxycholecalciferol to 1-25 dihydroxycholecalciferol.

**7. All of the following are the manifestation of hypocalcemic tetany EXCEPT**

**A.** Adduction of thumb

**B.** Flexion at metacarpophalyngial joint

**C.** Flexion at interphalyngial joint

**D.** Flexion at wrist

**Ans. 7. (C)**

There is hyperextension at interphalangeal joints. Due to the decreased calcium there is increased synaptic excitability.

**8. Word hormone means**

**A.** Chemical messenger
**B.** Neurotransmitter
**C.** To set in motion
**D.** Second messenger

**Ans. 8. (C)**

Word hormone was coined by Starling.

**9. A chemical substance acting on a neighboring cell is called**

**A.** Autocrine **B.** Paracrine
**C.** Juxtacrine **D.** Hormone

**Ans. 9. (B)**

**10. A chemical substance acting on a cell, which secretes it is called**

**A.** Autocrine **B.** Paracrine
**C.** Juxtacrine **D.** Hormone

**Ans. 10. (A)**

**11. A word hormone was coined by**

**A.** Bohr **B.** Starling
**C.** Sutherland **D.** Banting

**Ans. 11. (B)**

Hormone is a chemical substance secreted in the minute quantity by the ductless glands directly in to blood.

**12. All of the following are the functions of the endocrine gland EXCEPT**

**A.** Controls the magnitude of the biochemical reaction of target tissues
**B.** Controls circulating levels of energy yielding substances
**C.** Controls bodily process such as growth, metabolism, reproduction
**D.** Initiates the new functions by the cells

**Ans. 12. (D)**

Endocrine secretions can influence the existing functions of the cell.

**13. Hormone present in the bound form**

**A.** Represents the active form
**B.** Can easily be excreted by the kidney

C. Acts as a reserve form
D. Can be easily destroyed

**Ans. 13. (C)**

The hormone secreted by an endocrine gland is transported in the blood either in the free form or in the bound to plasma proteins. This binding decreases the excretion by the kidney. Though hormone can remain in circulation for a longer period it can bring about the effect in the free form.

**14. Which of the following hormone has the shortest half life**

A. Thyroid hormone B. Catecholamines
C. Aldosterone D. Cortisol

**Ans. 14. (B)**

Hormones belonging to the proteins and amines category are destroyed quickly in the body. Catecholamines belongs to amines category. Thyroid hormones, cortisol and aldosterone are bound to plasma proteins strongly than catecholamines.

**15. Which of the following hormone has the longest half life**

A. Thyroid hormone B. Adrenaline
C. Testosterone D. Cortisol

**Ans. 15. (A)**

**16. All of the following hormones use cAMP as a second messenger EXCEPT**

A. Glucagon
B. Thyroid stimulating hormone
C. Angiotensin II
D. Vasopressin

**Ans. 16. (C)**

Second messengers are the chemical substance which helps in relaying the information from the first messenger {hormones, voltage change} to the cellular machinery. cAMP is one of the widely used second messenger by many first messengers. The role of cAMP is widely studied in glucagon.

**17. Which of the following hormone uses calcium as a second messenger**

A. Parathyroid hormone
B. Calcitonin

C. Luteinizing hormone
D. Oxytocin

**Ans. 17. (D)**

Others uses cAMP as second messengers.

**18. Diacylglycerol, a 2nd messenger formed from the membrane phospholipid**

A. Activates protein kinase C in the cytoplasm
B. Causes release of calcium from endoplasmic reticulum
C. Activates membrane phospholipase C
D. Helps in the regulation of membrane proteins

**Ans. 18. (A)**

Diacylglycerol (DAG) is formed from the membrane phospholipid $PIP_2$ (phosphoinositolbiphosphate) by the action of membrane bound enzyme phospholipase C. This enzyme converts $PIP_2$ to Inositoltriphosphate ($IP_3$) and DAG.

**19. All of the following are tropic hormones EXCEPT**

A. Thyroid stimulating hormones
B. Prolactin
C. Adrenocorticotropic hormone
D. Luteinizing hormone

**Ans. 19. (B)**

Tropic hormones are the one which stimulates the secretion from other endocrine glands or tissues. TSH causes secretion of thyroid hormone from thyroid gland, LH causes secretion of testosterone from testes, ACTH affects the release of adrenal cortical hormones from adrenal glands.

**20. The acidophilic cells of anterior pituitary secretes**

A. Thyroid stimulating hormones
B. Follicle-stimulating hormone
C. Growth hormone
D. Adrenocorticotropic hormone

**Ans. 20. (C)**

Acidophilic cells secret GH and prolactin rest of the hormones are secreted by basophilic cells.

**21. The genes for the growth hormone are located on**

A. Chromosome 17 B. Chromosome 6

C. Chromosome 10 D. Chromosome 8

**Ans. 21. (A)**

**22. The genes for the prolactin hormone are located on**

A. Chromosome 17 B. Chromosome 6

C. Chromosome 10 D. Chromosome 8

**Ans. 22. (B)**

**23. Dwarfism in the Levi-Lorain dwarf results from**

A. The absence of growth hormone receptors

B. Presence of antibodies to the growth hormone receptors

C. Absence of high affinity GH binding proteins

D. Low level of growth hormone secretion

**Ans. 23. (C)**

Dwarfism does not always result from the absence of growth hormone. It can also result from decreased level of high affinity binding proteins (which results in increased degradation of the hormone (decreased half life ) or decreased level of somatomedin C as in Pygmies. In Levi - Lorain (Laron) dwarfs the number of growth hormone receptors are normal but there is decreased responsiveness.

**24. Growth hormone has all of the following effect on the protein metabolism EXCEPT**

A. Increased protein synthesis

B. Increased amino acid level in blood

C. Decreased urinary urea-nitrogen

D. Positive nitrogen balance

**Ans. 24. (B)**

GH is a protein anabolic hormone. It increases the formation of protein in the cell by increased protein synthesis, increased amino acid transport, decreased protein catabolism.

**25. Growth hormone**

A. Increases the utilization of the glucose for energy

B. Decreases glycogen deposition

**C.** Inhibits insulin secretion from beta cells
**D.** Initially increases the uptake of glucose by cells

**Ans. 25. (D)**

GH has a carbohydrate and protein sparer effect. This is primarily because of increased utilization of fat for energy. GH has a duak effect on uptake of glucose by cells, to begin with it has insulin like effect followed by anti-insulinic effect.

**26. Lipolytic effect of growth hormone is suppressed by**

**A.** Glucose administration **B.** Fasting
**C.** Stress **D.** Hypoglycemia

**Ans. 26. (A)**

All other condition results in low blood glucose level and or increased need for energy, and stimulates the secretion of GH and increases the lipolytic effect to provide energy.

**27. The effect of growth hormone on fat metabolism includes all of the following EXCEPT**

**A.** Breakdown of stored fat
**B.** Decreased conversion of fatty acids to acetyl CoA
**C.** Increased free fatty acid level in blood
**D.** Increased formation of acetoacetic acid

**Ans. 27. (B)**

GH has a catabolic effect on fat metabolism. This effect helps to spare the protein from being utilized for energy thus helps in the growth. It enhances the conversion of fatty acids to acetyl CoA which is than used for energy. Excessive breakdown of fat results in conversion of fat to acetoacetic acid by liver.

**28. Somatomedin C**

**A.** Promotes the uptake of thymidine in the bone
**B.** Activity is reduced in by protein deficiency
**C.** Activity is stimulated by cortisol
**D.** Has anti-insulinic activity

**Ans. 28. (A)**

Somatomedin C is also known as insulin-like growth factor-I (IGF-I). Most of its action are like insulin. It has a longer half life of about 20 hours as opposed to that of 20 minutes of GH. It helps in the cartilage growth by promoting the uptake of thymidine and sulphate by cartilage cells.

**29. All of the following decreases the secretion of somatomedin C EXCEPT**

**A.** Large doses of estrogen
**B.** Untreated diabetes mellitus
**C.** Growth hormone deficiency
**D.** Increased cortisol level

**Ans. 29. (D)**

Somatomedins are formed by the liver under the effect of GH. Though somatomedin C secretion is independent of GH level before birth after the birth its secretion is controlled by GH. The level of somatomedin C starts rising in childhood and peaks at the time of puberty and than falls in old age.

**30. Which of the following statement regarding insulin-like growth factor-II [IGF-II] is NOT TRUE**

**A.** IGF - II activity is controlled only by growth hormone
**B.** It plays important role in the growth of the foetus before birth
**C.** In adults genes for IGF - II are expressed in choroids plexus
**D.** The mRNAs for IGF - II are found specially in the liver and cartilage

**Ans. 30. (A)**

IGF-II secretion is independent of GH effect. It is responsible for the growth of the foetus before birth. If it is overexpressed there is disproportionate growth of few tissues, e.g. tongue, muscles, liver.

**31. Growth hormone increases the**

**A.** Absorption of calcium from GIT
**B.** Excretion of sodium by kidney
**C.** Excretion of potassium by kidney
**D.** Excretion of magnesium by kidney

**Ans. 31. (A)**

GH increases the absorption of calcium from GIT and sodium and potassium by renal tubules. This effect on kidney is not dependent on the adrenal gland secretion.

This absorbed minerals are diverted to the growing tissues.

**32. Growth hormone secretion is stimulated by**

**A.** Infusion of glucose
**B.** Protein deficient meal
**C.** Deep sleep
**D.** Increased level of cortisol

**Ans. 32. (C)**

The basal GH concentration ranges between 1.6-3 ng/ml. GH secretion is increased by condition which leads to hypoglycemia. Increase in glucose decreases the GH secretion. Amino acid especially arginine is known to stimulate GH secretion. REM sleep is associated with decreased GH level.

**33. Growth hormone secretion is inhibited by**

**A.** Glucagon
**B.** Estrogen
**C.** Androgens
**D.** Growth hormone

**Ans. 33. (D)**

The response to glucagon is been used as a test to check the GH reserve. Estrogen and androgens are shown to stimulate GH secretion.

**34. Person suffering from Acromegaly has all of the following features EXCEPT**

**A.** Increased hand and foot size
**B.** Prognathism
**C.** Wide spaced teeth
**D.** Dry skin

**Ans. 34. (D)**

Acromegaly results from the oversecretion of GH after the epiphysis have fused. The commonest cause is the acidophilic tumor of anterior pituitary. There is increase in the size of the acral bones (hands and feets) and the membranous bones. This causes increase in the hand and feet size, protrusion of the mandible and frontal bone with hypertrophy of the soft tissues.

**35. Receptors for adrenal cortical hormones are present**

**A.** On the cell membrane
**B.** In the cytoplasm

C. In the nucleus
D. None of the above

**Ans. 35. (B)**

Adrenal cortical hormones belong to the category of the steroid hormones. They can easily pass the lipid bilayer cell membrane.

**36. Glucocorticoid hormones**
A. Stimulates gluconeogeneis
B. Increases glucose utilization by the cells
C. Increases hepatic lipogenesis
D. Decreases glycogen storage

**Ans. 36. (A)**

Glucocorticoid hormones has protein and fat catabolic effect. It increases the blood glucose level by decreasing the uptake, utilization of the glucose by the cells and increasing the gluconeogenesis.

**37. Cortisol causes**
A. Increase in muscle proteins
B. Decrease in liver proteins
C. Decrease in plasma proteins
D. Increased amino acids in blood

**Ans. 37. (D)**

Cortisol causes breakdown of proteins in extrahepatic tissue. This causes release of amino acid in to blood and increased supply of amino acids to liver. Liver converts this to proteins and released into plasma.

**38. All of the following are the effect of cortisol on fat metabolism EXCEPT**
A. Mobilization of fat from adipose tissues
B. Increased level of free fatty acids in plasma
C. Decreased use of free fatty acid for energy
D. Increased ketone body formation

**Ans. 38. (C)**

Cortisol causes breakdown of fat in the adipose tissues. This causes release of free fatty acid in the blood which is taken to the liver and is also used for energy. Increased fatty acid supply to liver can get converted to ketone bodies especially in the absence of insulin.

**39. All of the following are present during inflammation EXCEPT**
A. Erythema
B. Infiltration by leucocytes

C. Swelling

D. Decreased temperature

**Ans. 39. (D)**

Inflammation is characterized by increased blood supply (erythema), increased temperature (calor), oozing of fluid from the blood vessels and infiltration of the WBC at the inflamed site (due to chemotaxis).

**40. Cortisol acts as an anti-inflammatory agent by**

A. Stabilizing lysosomal membrane

B. Increasing capillary permeability

C. Increasing chemotaxis

D. Increasing release of interleukin-I from WBCs

**Ans. 40. (A)**

The basic cause of inflammation is the breakdown of the lysosome and release of their contents to the surrounding. Interleukin - I is one of the mediator of inflammation. By stabilizing the lysosomal membrane cortisol makes the membrane more resistant to rupture and thus decreasing the release of the inflammatory mediator.

**41. All of the following statement about cortisol are true EXCEPT**

A. It causes hyperglycemia

B. It speeds up healing

C. It is a protein catabolic hormone

D. It depresses immunity

**Ans. 41. (B)**

Though it acts as an anti-inflammatory, it also inhibits the fibroblastic activity, which is needed for the healing process to occur. By breaking down the proteins and suppressing the production of lymphocytes it suppresses the immunity especially cell mediated immunity.

**42. Cortisol acts as anti-allergic by**

A. Blocking formation of antigen-antibody complex

B. Preventing formation of antibody

C. Blocks the effects of released histamine

D. Prevents the release of histamine by antigen-antibody complex

**Ans. 42. (D)**

Allergy is caused by release of histamine and few other allergic substance because of antigen-antibody interaction.

Cortisol does not have effect on the process of Ag-Ab binding or the histamine which is already released but it can prevent the release of histamine. Thus cortisol is of not much value in the acute effect of allergy but helps to prevent the subsequent attack.

**43. Cortisol increases the**
- **A.** Number of eosinophils in the blood
- **B.** Number of neutrophils in the blood
- **C.** Number of lymphocytes in the blood
- **D.** Number of monocytes in the blood

**Ans. 43. (B)**

All other decreases. The probable cause for increase neutrophilic count is the decreased immunity and hence increased chance to exposure to the infection.

**44. Cortisol decreases the**
- **A.** Number of red blood cells in the blood
- **B.** Number of platelets in the blood
- **C.** Number of lymphocytes in the blood
- **D.** Number of neutrophils the blood

**Ans. 44. (C)**

All other cell count increases.

**45. Cortisol plays a permissive role in all of the following EXCEPT**
- **A.** HCL secretion by stomach
- **B.** Vasopressor effect of catecholamines
- **C.** Lipolytic effect of catecholamines
- **D.** Lipolytic effect of glucagons

**Ans. 45. (A)**

Permissive effect means that the hormone itself cannot bring about the particular action but the presence of a hormone is must for some other chemical substance to have that particular effect.

**46. The conversion of angiotensin-I to angiotensin-II in lungs by angiotensin converting enzyme occurs in**
- **A.** Pulmonary capillary endothelium
- **B.** Alveolar cells
- **C.** Lung parenchyma
- **D.** Bronchus

**Ans. 46. (A)**

Angiotensin-I is formed by the action of rennin on plasma protein (angiotensinogen). This angiotensin-I is converted to angiotensin-II in the endothelium of the pulmonary capillaries. This angiotensin-II helps in the regulation of blood pressure.

**47. Aldosterone increases the reabsorption of**

**A.** Potassium by renal tubules
**B.** Calcium by renal tubules
**C.** Sodium by renal tubules
**D.** Hydrogen by renal tubules

**Ans. 47. (C)**

Aldosterone is the principal mineralocorticoid secreted by the zona glomerulosa of the adrenal cortex. It increases the plasma sodium level by reabsorbing sodium from renal tubules in exchange of potassium and hydrogen.

**48. Aldosterone increases the renal reabsorption of sodium from all of the following EXCEPT**

**A.** Distal convoluted tubules
**B.** Loop of Henle
**C.** Proximal convoluted tubules
**D.** Collecting tubules

**Ans. 48. (B)**

**49. Hyperaldosteronism is associated with**

**A.** Acidosis
**B.** Decreased urinary excretion of calcium
**C.** Hyperkalemia
**D.** Alkolosis

**Ans. 49. (D)**

Aldosterone causes reabsorption of sodium with excretion of hydrogen. This decreases the H ion level in body.

**50. Primary aldosteronism is associated with**

**A.** Muscle paralysis **B.** Cardiac toxicity
**C.** Decreased pH **D.** Edema

**Ans. 50. (A)**

Hypersecretion of aldosterone leads to increase excretion of potassium from kidney. This leads to hypokalemia which affects the RMP and results in hyperpolarisation making it more difficult for the tissue to be excited.

**51. The most potent factor [in terms of feedback gain] controlling the aldosterone secretion is**

**A.** Rennin angiotensin system
**B.** Potassium concentration in blood

C. ACTH

D. Sodium concentration in blood

**Ans. 51. (B)**

Though most important factor regulation aldosterone secretion is renin angiotensin system, the feedback gain effect of serum potassium is more than that of renin angiotensin system.

**52. The most important regulator of aldosterone secretion is**

A. Renin angiotensin system

B. Potassium concentration in blood

C. ACTH

D. Sodium concentration in blood

**Ans. 52. (A)**

Though the primary effect of aldosterone is to maintain normal sodium level, it is the renin angiotensin system is the main regulator of aldosterone secretion. ACTH plays more important role in the regulation of cortisol secretion.

**53. Renin secretion is stimulated by**

A. Sodium depletion in distal tubule

B. β- adrenergic antagonists

C. Parasympathetic stimulation

D. Increase in ECF volume

**Ans. 53. (A)**

Rest other factor inhibits the secretion of renin by juxtaglomerular cells.

**54. The intracellular effect of ACTH is mediated by**

A. Calcium  B. Diacylglycerol

C. Protein kinase C  D. Cyclic AMP

**Ans. 54. (C)**

**55. The intracellular effect of angiotensin-II is mediated by**

A. Inositol triphosphate  B. Diacylglycerol

C. Protein kinase A  D. Cyclic GMP

**Ans. 55. (B)**

**56. The intracellular effect of potassium is mediated by**

A. Calcium  B. Cyclic AMP

C. Protein kinase C  D. Cyclic GMP

**Ans. 56. (A)**

**57. The colloid present in the thyroid follicular cell is composed of**

**A.** Lipids
**B.** Carbohydrates
**C.** Phospholipid
**D.** Thyroglobulin

**Ans. 57. (D)**

Thyroid gland is unique in that the hormone synthesis occurs extracellularly at the colloid-cell interface. Thyroglobulin is composed of tyrosine amino acids which binds to iodine ion to form thyroid hormones.

**58. Inactive thyroid gland**

**A.** Has abundant colloid
**B.** Has small follicles
**C.** Lining cells are columnar
**D.** Shows the presence of resorption lacunae

**Ans. 58. (A)**

**59. The active thyroid gland shows all of the following characteristics EXCEPT**

**A.** Follicles are smaller in size
**B.** The lining cells are flat
**C.** Presence of resorption lacunae
**D.** Colloid is reduced

**Ans. 59. (B)**

When thyroid gland becomes active the protein in the colloid in which the thyroid hormones are present is broken down thus colloid material becomes less. The flat epithelial lining of the inactive thyroid follicles changes to columnar.

**60. The thyroid hormone synthesis takes place**

**A.** Inside the follicular cell
**B.** In the follicular lumen
**C.** At colloid-apex interface
**D.** At the basal surface of the follicular cells

**Ans. 60. (C)**

**61. For the normal synthesis of thyroid hormone, iodine requirement is**

**A.** 50 mg per year
**B.** 100 mg per year
**C.** 25 mg per year
**D.** 75 mg per year

**Ans. 61. (A)**

Iodine is the raw material required for the synthesis of thyroid hormones. Deficiency of iodine results in hypothyroidism. To prevent iodine as a limiting factor common salt is iodised, for one lakh part of NaCl one part of NaI is added. The weekly requirement is about 1 mg of Iodine.

**62. The other tissue beside thyroid where diiodotyrosine is formed is**

**A.** Salivary glands
**B.** Gastric mucosa
**C.** Ciliary body
**D.** Mammary glands

**Ans. 62. (D)**

But the significance of the formation of diiodotyrosine in mammary gland is not clear.

**63. After the synthesis of thyroid hormone, each thyroglobulin molecule contains**

**A.** 2-3 molecules of thyroxine
**B.** 5-6 molecules of thyroxine
**C.** 8-9 molecules of thyroxine
**D.** 12-13 molecules of thyroxine

**Ans. 63. (B)**

**64. After the synthesis of thyroid hormone, for every molecule of triiodothyronine there are**

**A.** 10-12 molecules of thyroxine
**B.** 7-9 molecules of thyroxine
**C.** 17-19 molecules of thyroxine
**D.** 22-24 molecules of thyroxine

**Ans. 64. (C)**

**65. Which of the following statement about transport of thyroxine hormone in the blood is TRUE**

**A.** Most of the thyroid hormone is transported in the free form
**B.** Most of the thyroxine is transported by globulin
**C.** Albumin has the lowest binding capacity for thyroxine
**D.** Most of the thyroxine is transported by prealbumin

**Ans. 65. (B)**

Thyroxine binding globulin has highest affinity for thyroxine while thyroxine binding albumin has the highest capacity for binding with thyroxine. About 98-99 % of the thyroid hormone is transported in bound form.

**66. The receptors for thyroid hormone are present**
**A.** On the cell membrane
**B.** In the cytoplasm
**C.** In the nucleus
**D.** None of the above

**Ans. 66. (C)**

Thyroid hormones belong to the category of the amine hormones. So it the receptor should be present on the cell membrane but the presence of iodine changes this and the hormones can easily cross the cell membrane.

**67. The magnitude of the calorigenic action of thyroid hormone depends on**
**A.** Level of the thyroid hormone
**B.** Level of TSH
**C.** Basal level of catecholamine and BMR
**D.** Rate of secretion of thyroid hormone

**Ans. 67. (C)**

The calorigenic action is the increased heat production. This depends on the level of catecholamines and BMR. If the catecholamine level and BMR is high the further increased thyroid level would not have much effect on the calorigenic action of thyroid hormones.

**68. Increase thyroid hormone level is associates with**
**A.** Decreased cholesterol level in blood
**B.** Increased phospholipid level in blood
**C.** Increased triglyceride level in blood
**D.** Decreased free fatty acid level in blood

**Ans. 68. (A)**

Thyroid hormone increases the number of LDL receptors on the membrane of the liver cells. This causes increase binding and excretion of LDL. LDL has high proportion of cholesterol.

**69. Hypothyroidism often results in**
**A.** Microcytic hypochromic anaemia
**B.** Normocytic normochromic anaemia
**C.** Normocytic hypochromic anaemia
**D.** Macrocytic hypochromic anaemia

**Ans. 69. (B)**

RBC count in blood is directly proportional to the BMR of the body. Hypothroidism is associated with decreased BMR which can result in decreased erythropoiesis.

**70. All of the following are seen in hyperthyroidism EXCEPT**

**A.** Muscle weakness
**B.** Tachycardia
**C.** Weight gain
**D.** Wide pulse pressure

**Ans. 70. (C)**

Hyperthyroidism is associated with increased appetite. Increased appetite should cause increase weight gain, but patients with hyperthyroidism loose weight. This could be because of (i) Increased metabolism which requires more energy (ii) Protein catabolism especially of muscle.

**71. Which of the following is seen in hypothyroidism**

**A.** Increased appetite **B.** Tachycardia
**C.** Clonus **D.** Menorrhagia

**Ans. 71. (D)**

All other are the features of hyperthyroidism. Menorrhagia is means excessive menstrual bleeding, hyperthyroidism is associated with oligomenorrhea.

**72. Insulin is secreted by**

**A.** Alpha cells of islets of Langerhans
**B.** Beta cells of islets of Langerhans
**C.** Delta cells of islets of Langerhans
**D.** PP cells of islets of Langerhans

**Ans. 72. (B)**

Beta cell forms the main bulk of the islet of Langerhans of pancreas. It occupies the central portion of the islets.

**73. Glucagon is secreted by**

**A.** Alpha cells of islets of Langerhans
**B.** Beta cells of islets of Langerhans
**C.** Delta cells of islets of Langerhans
**D.** PP cells of islets of Langerhans

**Ans. 73. (A)**

Alpha cells are the next abundant cells in the islets of Langerhans usually occupies the peripheral part of the islets.

**74. Delta cells of islets of Langerhans secrets**

A. Insulin
B. Pancreatic polypeptide
C. Secretin
D. Somatostatin

**Ans. 74. (D)**

Somatostatin is also found in brain where it inhibits the secretion of growth hormone from anterior pituitary.

**75. F cells present in islets of Langerhans secrets**

A. Cholecystokinin
B. Pancreatic polypeptide
C. Secretin
D. Somatostatin

**Ans. 75. (B)**

F cells are also known as PP (pancreatic polypeptide) cells.

**76. The central mass of the islets of Langerhans is constituted by**

A. Alpha cells B. Beta cells
C. Delta cells D. PP cells

**Ans. 76. (B)**

**77. The major cell type present in the islets of Langerhans is**

A. Alpha cells B. Beta cells
C. Delta cells D. PP cells

**Ans. 77. (B)**

**78. The gene for insulin is located on**

A. Chromosome 11 B. Chromosome 5
C. Chromosome 20 D. Chromosome 1

**Ans. 78. (A)**

**79. The functional index of beta cells in person receiving exogenous insulin is provided by**

A. C peptide level in blood
B. Proinsulin level
C. Prekallikrein level
D. Insulin level

**Ans. 79. (A)**

Insulin is synthesized as preproinsulin which than splits to form proinsulin. Here insulin is bound to C (connecting) peptide. The insulin is formed with the separation from C peptide. When insulin is released from beta cell the C peptide is also secreted along with it. To know the functional activity of the beta cells in diabetes mellitus the level of C peptide is measured.

**80. Normal basal rate of insulin release from beta cells is**

**A.** 10 IU / hour **B.** 100 IU / hour
**C.** 1 IU / hour **D.** 50 IU / hour

**Ans. 80. (C)**

Insulin level is determined by the blood glucose level. It can very from very low secretion to 50 IU / hour in response to glucose intake. On an average the rate is about 1 IU / hour.

**81. The gene for insulin receptor is located on**

**A.** Chromosome 19 **B.** Chromosome 5
**C.** Chromosome 10 **D.** Chromosome 11

**Ans. 81. (A)**

**82. All of the following insulin effects occur very fast EXCEPT**

**A.** Increased transport of glucose in cells
**B.** Increased transport of amino acids in cells
**C.** Increased transport of potassium in cells
**D.** Stimulation of protein synthesis

**Ans. 82. (D)**

Some of the actions of insulin occur within in a span of few seconds (rapid actions), some take about minutes to occur (intermediate actions), while some take a longer time about hours to occur (delayed actions). Rapid action includes increased transport of amino acid, potassium, and glucose into insulin sensitive cells. Stimulation of protein synthesis occurs few minutes after release of insulin.

**83. Insulin lowers the blood glucose level by**

**A.** Decreased rate of glucose release from liver
**B.** Inhibiting glycogen synthesis
**C.** Increasing plasma free fatty acid level
**D.** By inhibiting glycolysis

**Ans. 83. (A)**

Insulin increases the uptake of glucose by hepatocytes. If excessive glucose enters the hepatic cell the glucose is stored as glycogen (up to 6% of liver mass), and excess of glucose is converted to fats (very LDL) and transported to adipose tissue.

**84. Insulin promotes hepatic uptake, storage and use of glucose by**

**A.** Decreasing the activity of phosphofructokinase
**B.** Stimulating the activity of glucokinase
**C.** Stimulating activity of glycogen phosphorylase
**D.** Stimulating glycogenolysis

**Ans. 84. (B)**

Glucokinase is the enzyme which brings about the initial phosphorylation of glucose. This phosphorylated glucose gets trapped inside the cell.

**85. The muscle glycogen cannot be converted back to glucose because of absence of enzyme**

**A.** Glucose phosphatase
**B.** Glucokinase
**C.** Glucose-6-phosphatase
**D.** Phosphofructokinase

**Ans. 85. (A)**

Glucose phosphate converts phosphorylated glucose to free glucose. This enzyme is absent in muscle cell and hence glycogen cannot be converted back to glucose.

**86. Insulin increases the transport of glucose across muscle cell by stimulating**

**A.** Glucose transporter 1
**B.** Glucose transporter 2
**C.** Glucose transporter 5
**D.** Glucose transporter 4

**Ans. 86. (D)**

There are seven different types of GLUT transporters namely GLUT-1 to GLUT-7. Insulin sensitive cells have GLUT-4 transporter containing vesicles sensitive to exercise and insulin. When insulin activates the cell these vesicles move to the membrane and insert the transporter and when insulin activity stops these transporters are endocytosed.

**87. The basal glucose uptake by placenta, brain, RBCs occurs through**

**A.** Glucose transporter 1
**B.** Glucose transporter 2

C. Glucose transporter 5
D. Glucose transporter 4

**Ans. 87. (A)**

Brain, placenta, RBC does not depend on presence of insulin. In these tissue basal glucose uptake occurs through GLUT-1 and GLUT-3.

**88. Insulin facilitates the uptake of glucose in**

A. Intestinal mucosa B. Renal tubules
C. Adipose tissue D. RBCs

**Ans. 88. (C)**

Facilitated diffusion of glucose in intestinal mucosa, renal tubules and RBC is independent of insulin.

**89. Insulin enhances the transport of glucose across the membrane of all of the following tissue EXCEPT**

A. Cardiac muscle B. Leukocytes
C. Fibroblasts D. Renal tubules

**Ans. 89. (D)**

**90. Insulin enhances the transport of glucose across the membrane of all of the following tissue EXCEPT**

A. Smooth muscle B. Lens of the eye
C. Pituitary D. Intestinal mucosa

**Ans. 90. (D)**

**91. Insulin enhances the transport of glucose across the membrane of all of the following tissue EXCEPT**

A. Mammary gland
B. Brain
C. Aorta
D. Alpha cells of pancreas

**Ans. 91. (B)**

**92. Which of the following statement about the effect of insulin on fat metabolism is TRUE**

A. Most of the lipogenesis occurs in adipose tissue
B. Insulin increases the activity of enzyme hormone sensitive lipase
C. Decreases the activity of enzyme lipoprotein lipase

**D.** Isocitrate ions increases the activity of acetyl CoA carboxylase

**Ans. 92. (D)**

The excessive glucose that enters the liver (after the saturation for the glycogen storage has reached the) gets converted to triglycerides and in this form fat is transported to adipose tissue. The first step in the lip [ogenesis is conversion of acetyl CoA to malonyl CoA. This conversion is brought by the enzyme acetyl CoA carboxylase. The activity of this enzyme is increased by citrate and isocitrate ions.

**93. Insulin has all of the following effects on adipose tissue EXCEPT**

**A.** Decreased glycerol phosphate synthesis
**B.** Increased triglyceride deposition
**C.** Increased potassium uptake
**D.** Inhibition of hormone sensitive lipase enzyme

**Ans. 93. (A)**

Insulin increases the activity of lipoprotein lipase enzyme present in the capillary wall of the adipose tissue. This enzyme breaks the triglycerides in the blood to glucose and FFA. The fats can not enter the cell in triglycerides form. Insulin increases the uptake of glucose by adipose cells. This glucose is used to form alpha-glycerophosphate. The alpha-glycerophosphate and FFA combine again to form triglycerides inside the adipose cells.

**94. All of following statement regarding effect of insulin on muscle are true EXCEPT**

**A.** Increased amino acid uptake
**B.** Increased ketone uptake
**C.** Increased potassium uptake
**D.** Increased release of gluconeogenic amino acids

**Ans. 94. (D)**

Insulin acts as fat and protein sparer hormone as it facilitates the use of glucose for energy. It inhibits gluconeogenesis. Insulin is a protein anabolic hormone and thus prevents the breakdown of protein with little release of free amino acid.

**95. The hepatic effect of insulin includes all of the following EXCEPT**

**A.** Increased protein synthesis
**B.** Decreased gluconeogenesis

**C.** Increased ketogenesis
**D.** Increased glycogen synthesis

**Ans. 95. (C)**

In the absence of insulin, the predominantly used substance for energy is fat. This results in formation of excessive amount of FFA. In the liver in the absence insulin there is increased transport of this FFA by carnitine to mitochondria. The beta oxidation of FFA in mitochondria results in the formation of acetoacetic acid. The decreased utilization of this acetoacetic acid by the body results in conversion of some of this acetoacetic acid in to beta-hydroxybutyric acid and acetone.

**96. Insulin secretion is stimulated by**
**A.** Leucine
**B.** Alpha deoxyglucose
**C.** Epinephrine
**D.** Beta-adrenergic blockers

**Ans. 96. (A)**

Insulin increases the transport of amino acids in to cells and also the release of insulin is stimulated by amino acids. But the amino acids involved are different.

**97. Insulin secretion is inhibited by**
**A.** Glucose
**B.** Potassium depletion
**C.** Glucagon
**D.** Acetylcholine

**Ans. 97. (B)**

Rest of the factor are known to increase insulin secretion.

**98. All of the following factors increase the insulin secretion EXCEPT**
**A.** Gastrin
**B.** Beta-adrenergic stimulators
**C.** Exercise
**D.** Decreased magnesium in blood

**Ans. 98. (C)**

During exercise muscle requires rapid supply of energy, which can be supplied with glucose. The increased glucose transport is brought about by GLUT $ transporter sensitive to exercise.

**99. All of the following factors inhibit the insulin secretion EXCEPT**

A. Beta - ketoacids
B. Somatostatin
C. Serotonin
D. Alpha - adrenergic stimulators

**Ans. 99. (A)**

**100. Which of the following hormone inhibits the insulin secretion**

A. Growth hormone  B. Glucagon
C. Cortisol  D. Insulin

**Ans. 100. (D)**

**101. All of the following are usually associated with type I diabetes mellitus EXCEPT**

A. Genetic predisposition
B. Presence of antibody
C. Obesity
D. Insulin dependence

**Ans. 101. (C)**

Type I diabetes is also known as IDDM (insulin-dependent diabetes mellitus, juvenile DM). It is common in the younger individual and in those having a genetic background of DM. Commonest abnormality is presence of anti–beta cell antibody in plasma. It is primarily a T-lymphocyte mediated disease. It is associated with high incidence of acidosis and ketosis.

**102. All of the following are usually associated with type II diabetes mellitus EXCEPT**

A. Resistance to insulin action
B. Obesity
C. Normal insulin level
D. Young age

**Ans. 102. (D)**

Type II diabetes is also known as NIDDM (non-insulin-dependent diabetes mellitus). It is associated either with normal level of insulin secretion or decreased level of insulin secretion. This is common in older age group (after 40 years and obese people. It is the most common type of diabetes. In this the insulin receptors are resistant to the action of insulin. It also has the genetic component.

**103. Diabetes mellitus is associated with**

**A.** Increased weight gain
**B.** Decreased plasma osmolality
**C.** Increased hunger
**D.** Hypokalemia

**Ans. 103. (C)**

Basic symptoms of DM includes polydypsia, polyphagia, polyurea and non-healing ulcer. Insulin lack is associated with intracellular depletion of potassium but frequently the extracellular potassium is high. Increased blood glucose increases the osmolality of plasma.

**104. Antidiuretic secretion is increased by**

**A.** Alcohol
**B.** Decreased effective osmotic pressure
**C.** Increased ECF volume
**D.** Standing

**Ans. 104. (D)**

ADH is also has vasoconstrictor ability and so also called as AVP (arginine vasopressin). Increased osmolar pressure (and decreased ECF volume through volume receptors) stimulates the neurons in the hypothalamus to release more ADH so as to conserve water and bring osmotic pressure back to the normal. Standing is associated with the fall in blood pressure to counter this body releases more AVP to bring about vasoconstriction and water conservation.

**105. Actions of oxytocin includes all of the following EXCEPT**

**A.** Contraction of pregnant uterus
**B.** Synthesis of milk
**C.** Ejection of milk
**D.** Contraction of vas deferens

**Ans. 105. (B)**

Oxytose is a substance which can cause contraction of the pregnant uterus. In male it causes contraction of vas deferens, thus facilitating the transport of sperm. The synthesis of the milk is a function of prolactin.

**106. Probably the only endocrine gland with true secretomotor innervation is**

**A.** Pituitary  **B.** Thyroid
**C.** Adrenal medulla  **D.** Pancreas

**Ans. 106. (C)**

The secretion of the other endocrine glands is controlled by hormones. Adrenal medulla being apart of the ANS releases adrenaline and noradrenaline in response to ANS stimulation.

**107. Noradrenalin is converted to adrenalin by the enzyme**

**A.** Dopamine β-hydroxylase
**B.** Phenylethanalamine-N-Methyltransferase
**C.** DOPA decarboxylase
**D.** Phenylalanine hydroxylase

**Ans. 107. (B)**

Norepinephrine is methylated with the help of enzyme PNMT to form epinephrine. About 80% of NE is converted to epinephrine in adrenal medulla.

**108. During the synthesis of adrenalin, enzyme tyrosine hydroxylase helps in the conversion of**

**A.** Tyrosine from phenylalanine
**B.** Dihydroxyphenylalanine to dopamine
**C.** Tyrosine to dihydroxyphenylalanine
**D.** Noradrenaline to adrenaline

**Ans. 108. (C)**

Conversion of tyrosine to DOPA (dihydroxyphenylalanine) is the first step in the synthesis of adrenalin. The tyrosine is hydroxylated to form DOPA with the help of enzyme tyrosine hydroxylase.

**109. The quantitatively the most important means of removing noradrenalin from the circulation is by**

**A.** Its reuptake in nerve terminals
**B.** The action of enzyme COMT
**C.** Its methylation
**D.** Its oxidation

**Ans. 109. (A)**

NE reuptake occurs by active transport and this accounts for removal of 50-80 % of the secreted NE. The remaining NE is removed mostly by diffusion to the surrounding fluids and to a very little extent it is destroyed by the enzymes COMT and MAO.

**110. The enzyme catechol-O-methyltransferase [COMT] is abundant in all of the following organs EXCEPT**

A. Brain B. Liver

C. Kidney D. Nerve endings

**Ans. 110. (D)**

MAO is present in the nerve endings.

**111. The ionotropic and chronotropic effect of catecholamine is mediated via**

A. $\alpha$-1 adrenergic receptors

B. $\alpha$-2 adrenergic receptors

C. $\beta$-1 adrenergic receptors

D. $\beta$-2 adrenergic receptors

**Ans. 111. (C)**

Beta-2 and Alpha receptors are present in the vascular wall. Alpha receptor causes vasoconstriction while beta-2 receptors bring about vasodilatation. Heart has predominantly beta-1 receptors.

# 12

# GIT MCQs

**1. Secretory immunity in the gastrointestinal tract is provided by**

A. IgG  B. IgA
C. IgM  D. IgD

**Ans. 1. (B)**

IgA is secreted in external secretions of the gastro-intestinal tract, respiratory tract, and genitourinary system, and in tears, saliva and colostrum. Secretory IgA has antibody activity to bacterial and viral antigens, toxins and dietary macromolecules.

**2. Auerbach's plexus of the wall of the stomach is present**

**A.** In the serosa
**B.** Between the middle circular muscle layer and the mucosa
**C.** Between the outer longitudinal muscle layer and middle circular muscle layer
**D.** In the muscularis mucosa

**Ans. 2. (C)**

Auerbach's plexus (myenteric plexus) is a network of nerve fibers present between outer longitudinal and middle circular muscle layers in the wall of the stomach, small intestine and colon. This, along with another neural plexus called, Submucous plexus (Meissner's plexus) that is present between middle circular muscle layer and mucosa, constitute the enteric nervous system of the gut.

**3. Which of the following is NOT TRUE as regards peristalsis?**

**A.** Peristalsis is a reflex response
**B.** It occurs in all parts of the GIT except the oesophagus

**C.** Its occurrence is independent of the extrinsic innervation.
**D.** It is initiated when the gut wall is stretched by the contents of the lumen

**Ans. 3. (B)**

Peristalsis is a reflex response initiated when the gut wall is stretched. It occurs in all parts of gastrointestinal tract from the oesophagus to the rectum. Peristaltic activity can be increased or decreased by the autonomic input to the gut, but its occurrence is independent of extrinsic innervation.

**4. At low flow rate, saliva in the mouth is,**
**A.** Isotonic, acidic and low in $K^+$
**B.** Hypotonic, acidic and low in $K^+$
**C.** Hypertonic, alkaline and rich in $Na^+$
**D.** Hypotonic, slightly acidic and rich in $K^+$

**Ans. 4. (D)**

Saliva secreted in the acini is isotonic with concentrations of $Na^+$, $K^+$, $Cl^-$, $HCO_3^-$ that are close to those in plasma. The excretory ducts and intercalated ducts modify the composition of saliva by extracting $Na^+$, $Cl^-$ and adding $K^+$ and $HCO_3^-$. The ducts are relatively impermeable to water. Therefore at low salivary flows, the saliva that reaches mouth is hypotonic, slightly acidic and rich in $K^+$.

**5. The gastric motility is stimulated by**
**A.** Cholecystokinin
**B.** Distention of the stomach
**C.** Secretin
**D.** Peptide YY

**Ans. 5. (B)**

There are three pairs of salivary glands in humans. Parotid glands containing serous cells, Sublingual glands containing mucous cells and submandibular (submaxillary) glands containing both serous and mucous cells.

**6. Histamine stimulates acid secretion by binding with following receptor on parietal cells**
**A.** CCK-B receptors **B.** $H_2$ receptors
**C.** $H_1$ receptors **D.** $M_3$ receptors

**Ans. 6. (B)**

Ptyalin is an enzyme, also known as salivary α-amylase. Other three factors have antibacterial actions. Lysozyme attacks wall of bacteria, IgA is a secretory immunoglobulin, and Lactoferrin is bacteriostatic.

**7. The parotid gland is made up of**

A. Mucous cells
B. Serous cells
C. Mucous cells and serous cells
D. G cells

**Ans. 7. (B)**

Cholecystokinin, secretin and peptide YY are all inhibitors of gastric motility. Distention of stomach causes stretching of the stomach wall which elicit myenteric reflexes in the wall that excite the activity of pyloric pump and slightly inhibit the pylorus.

**8. Pepsinogen is converted to pepsin by**

A. Gastrin B. HCL
C. Secretin D. Somatostatin

**Ans 8. (B)**

Oxyntic (Parietal) cells of the oxyntic glands in the fundus and corpus of the stomach secrete HCL.

**9. HCL secretion is inhibited by**

A. Gastrin B. Acetylcholine
C. Somatostatin D. Histamine

**Ans. 9. (C)**

**10. The antibacterial factors in salivary secretion does not include**

A. Lactoferrin B. IgA
C. Ptyalin D. Lysozyme

**Ans. 10. (C)**

Histamine binds to $H_2$ receptors of the parietal cells, and via Gs, this increases adenyl cyclase activity and intracellular cAMP. Cyclic AMP acts via protein kinase A to increase the transport of $H^+$ into the gastric lumen by $H^+$- $K^+$ ATPase activity.

**11. Somatostatin is secreted by**

A. Delta cells
B. Enterochromaffin-like cells
C. Parietal cells
D. Chief cells

**Ans. 11. (A)**

Somatostatin is secreted by the D cells of the gastric mucosa. It acts as a paracrine agent and inhibits gastrin secretion by G cells and HCL secretion by parietal cells.

**12. Gastrin is secreted by**

A. Parietal cells  B. Delta cells
C. G cells  D. S cells

**Ans. 12. (C)**

Gastrin is secreted by G cells. G cells are present in the pyloric glands located in the antrum and pyloric region of the stomach. Gastrin released from the G cells, enters circulation and travels to orad stomach where it stimulates parietal cell HCL secretion.

**13. Major stimulus that causes release of secretin is**

A. Gastrin
B. Cholecystokinin
C. Protein digestion products
D. HCL

**Ans. 13. (D)**

Secretin is secreted by the S cells of duodenum. The major stimulus for secretin release is acidic pH in duodenum resulting from HCL in the chyme, with a pH of 4.5 being the threshold for S cell activation and maximal secretin release being achieved at pH = 3.0. Protein digestion products stimulate both secretin and cholecystokinin (CCK) secretion but are the major stimulants for CCK secretion. CCK potentiates the effects of secretin but does not affect its release.

**14. HCL is secreted by**

A. G cells  B. Chief cells
C. Delta cells  D. Oxyntic cells

**Ans. 14. (D)**

Pepsinogens are precursors of proteolytic enzyme pepsin. Pepsinogens are secreted by the chief cells (peptic cells) and mucous cells of gastric glands. They are activated by gastric hydrochloric acid to form active pepsin. Pepsin has a pH maximum of 1.6 to 3.2.

**15. Intrinsic factor is secreted by**

A. Chief cells
B. Parietal cells

C. Delta cells
D. Enterochromaffin-like cells

**Ans. 15. (B)**
Oxyntic (Parietal) cells of the oxyntic glands in the fundus and corpus of the stomach also secrete intrinsic factor along with HCL.

**16. Pepsin is**
A. Fat digesting enzyme
B. Secreted by oxyntic cells
C. Proteolytic enzyme
D. Milk clotting enzyme

**Ans. 16. (C)**
Pepsin, the active form of pepsinogen, is a proteolytic enzyme which hydrolyses the bonds between aromatic amino acids such as phenylalanine or tyrosine and a second amino acid.

**17. One of the following is exopeptidase**
A. Elastase
B. Carboxypeptidase
C. Trypsin
D. Chymotrypsin

**Ans. 17. (B)**
Trypsin, chymotrypsin and elastase act at interior peptide bonds in the peptide molecules and are called endopeptidases. The carboxypeptidases of the pancreas are exopeptidases that hydrolyze the amino acids at the carboxyl and amino ends of the polypeptides.

**18. Cholecystokinin is synthesized and released by**
A. S cells　　B. G cells
C. D cells　　D. I cells

**Ans. 18. (D)**
Cholecystokinin is synthesized and released from I cells of the duodenum.

**19. Secretin is synthesized and released by**
A. S cells of duodenum
B. ECL cells of stomach
C. D cells of stomach
D. I cells of duodenum

**Ans. 19. (A)**
Secretin is synthesized and released from S cells of the duodenum.

**20. Trypsinogen is**

A. Exopeptidase
B. Fat emulsifying agent
C. Secreted by D cells of stomach
D. Converted to trypsin by enterokinase

**Ans. 20. (D)**

Trypsinogen, an inactive proenzyme is converted to active enzyme, trypsin by the brush border enzyme enterokinase also known as enteropeptidase (which is secreted by the intestinal mucosa) when the pancreatic juice enters duodenum.

**21. Chymotrypsinogen is**

A. Converted to chymotrypsin by enterokinase
B. Exopeptidase
C. Converted to chymotrypsin by trypsin
D. Fat emulsifying agent

**Ans. 21. (C)**

Chymotrypsinogen is converted to chymotrypsin by trypsin. Trypsin also converts other proenzymes into their active enzymes, for example, it converts procarboxypeptidase into carboxypeptidase and proelastase into elastase.

**22. Pancreatic secretion is stimulated by**

A. Cholecystokinin
B. Secretin
C. Vagal stimulation
D. All of the above factors

**Ans. 22. (D)**

Secretin acts on pancreatic ducts to cause secretion of alkaline pancreatic juice which is rich in $HCO_3^-$. Cholecystokinin (CCK) acts on acinar cells to cause production of pancreatic juice rich in enzymes. Like CCK, acetylcholine also acts on acinar cells to cause discharge of zymogen granules, therefore vagal stimulation causes secretion of small amounts of pancreatic juice rich in enzymes.

**23. Increased circulating levels of serum bilirubin is known as**

A. Steatorrhea
B. Cholelithiasis
C. Jaundice
D. Sprue

**Ans. 23. (C)**

When free or conjugated bilirubin accumulates in the blood, the skin, scleras, and mucous membranes turn yellow. This yellowness is known as jaundice (icterus).

**24. Lithocholic acid is a**

A. Primary bile acid B. Bile salt
C. Secondary bile acid D. Bile pigment

**Ans. 24. (C)**

There are two major primary bile acids secreted by hepatic exocrine tissue namely, Cholic acid and Chenodeoxycholic acid. When they enter small intestine they are converted to secondary bile acids namely, Deoxycholic acids and Lithocholic acid.

**25. Primary bile acids are converted to secondary bile acids in**

A. Liver B. Gallbladder
C. Small intestine D. Pancreas

**Ans 25. (C)**

When primary bile acids enter the small intestine, they are dehydroxylated in the 7 position to either deoxycholic acid or lithocholic acid, by intestinal flora. These molecules are called secondary bile acids.

**26. Presence of Gallstones is known as**

A. Cholelithiasis B. Steatorrhea
C. Zollinger-Ellison syndrome
D. None of the above

**Ans. 26. (A)**

Presence of gallstones is known as cholelithiasis. These are of two types, calcium bilirubinate stones and cholesterol stones.

**27. Steatorrhea is caused due to**

A. Malabsorption of carbohydrates
B. Iron deficiency
C. Malabsorption of fats
D. Vitamin $B_{12}$ deficiency

**Ans. 27. (C)**

Patients with diseases that destroy the exocrine portion of pancreas have fatty, bulky and clay-coloured stools because of impaired digestion and absorption of fat. This condition is known as steatorrhea. Another cause of steatorrhea is defective reabsorption of bile salts in the distal ileum.

**28. Mucous in the small intestine is secreted by**

**A.** S cells

**B.** Brunner's glands in the duodenum

**C.** Delta cells

**D.** Peptic cells

**Ans. 28. (B)**

**29. Deficiency of intrinsic factor is associated with**

**A.** Steatorrhea

**B.** Pernicious anemia

**C.** Jaundice

**D.** Protein energy malnutrition.

**Ans 29. (B)**

For the absorption of vitamin $B_{12}$, intrinsic factor secreted by the parietal cells of gastric mucosa is required. In pernicious anemia there is autoimmune destruction of parietal cells leading to deficiency of vitamin $B_{12}$ as intrinsic factor secretion is affected.

**30. Peristaltic wave called mass movement is seen in**

**A.** Large intestine

**B.** Stomach

**C.** Small intestine

**D.** Oesophagus

**Ans. 30. (A)**

Peristaltic wave called mass movement occurs in the large intestine, in which there is simultaneous contraction of smooth muscle over large confluent areas. It serves to propel the luminal material toward the rectum.

# 13

# Nutrition, Digestion and Absorption

**1. Fructose is absorbed by**

**A.** Active transport

**B.** Facilitated diffusion

**C.** Osmosis

**D.** Simple diffusion

**Ans. 1. (B)**

Fructose absorption is independent of $Na^+$ or transport of glucose and galactose; it is transported instead by facilitated diffusion from the intestinal lumen into the enterocytes and out of enterocytes into interstitium.

**2. Emulsification of dietary lipids is brought about by**

**A.** Secretin
**B.** Cholecystokinin
**C.** Bile salts
**D.** Pepsin

**Ans. 2. (C)**

Bile salts are potent emulsifying agents of lipids. Fats are finely emulsified in small intestine by the detergent action of bile salts, lecithin and monoglycerides.

**3. Galactose absorption is by**

**A.** Na–dependent active transport process

**B.** Simple diffusion

**C.** Facilitated diffusion

**D.** Pinocytosis

**Ans. 3. (A)**

Like glucose, galactose is also actively transported from small intestinal lumen into the tissues by sodium-coupled carrier mechanism that is driven by the sodium concentration gradient established by the $Na^+$, $K^+$- ATPase localized at the basolateral membrane of the enterocytes.

**4. Lactose intolerance results from**

**A.** Lactase deficiency

**B.** Inadequate supply of milk in the diet

C. Fructose deficiency
D. Protein malabsorption

**Ans. 4. (A)**

Lactase is milk digesting brush border enzyme which is quite high during the neonatal period and through infancy. It begins to decline sharply after six years of age. In adults due to the deficiency of this enzyme there is intolerance to milk (lactose intolerance) which is characterized by diarrhea, bloating and flatulence.

**5. Gastric pepsin helps in digestion of**

A. Carbohydrate
B. Lipids
C. Iron
D. Proteins

**Ans. 5. (D)**

Pepsin hydrolyses the bonds between aromatic amino acids such as phenylalanine or tyrosine and a second amino acid, so the products of peptic digestion are polypeptides of diverse sizes.

**6. Lipid digestion in stomach is mainly brought about by**

A. Lingual lipase
B. Gastric lipase
C. Pancreatic lipase
D. Colipase

**Ans. 6. (A)**

Lingual lipase secreted by ebner's glands on the dorsal surface of the tongue is active in stomach and can digest as much as 30% of dietary triglycerides. Gastric lipase is of little importance except in pancreatic insufficiency. Most fat digestion, however, begins in the duodenum and pancreatic lipase is most important enzyme involved.

**7. Which of following statements is NOT TRUE regarding chymosin**

A. It is a milk clotting enzyme
B. It is also known as rennin
C. It is found in the stomachs of young animals
D. It is a pancreatic enzyme

**Ans. 7. (D)**

Chymosin is a milk clotting gastric enzyme which is also known as rennin. It is present in the stomachs of young animals but is probably absent in human.

**8. Absorption of bacterial antigens is by**

A. Brunner's glands of small intestine
B. Neck mucous cells of stomach

C. Microfold cells of intestine
D. Crypts of Lieberkühn

**Ans. 8. (C)**

Absorption of protein antigens, especially bacterial and viral proteins, takes place in large microfold cells or M cells, which are specialized intestinal epithelial cells that overlie the Peyer's patches. These cells pass the antigen to the lymphoid cells, and lymphocytes are activated.

**9. Short-chain fatty acids are produced in and absorbed from**

A. Large intestine
B. Small intestine
C. Stomach
D. Oesophagus

**Ans. 9. (A)**

Short-chain fatty acids are two or five-carbon weak acids which are produced in the colon and absorbed from it. They are formed by the action of colonic bacteria on complex carbohydrates, resistant starches and other components of dietary fiber, i.e., the material that escapes digestion in the upper gastrointestinal tract and enters colon.

**10. Protein digestion begins in**

A. Mouth
B. Oesophagus
C. Stomach
D. Small intestine

**Ans. 10. (C)**

Protein digestion does not occur in the mouth and oesophagus. It begins in the stomach, where pepsin breaks some of the peptide linkages.

**11. Any protein that appears in the stool derives from**

A. Diet
B. Partially digested proteins
C. Bacteria and cellular debris
D. Liver proteins

**Ans. 11. (C)**

Most of the digestion and absorption of the proteins occur in small intestine. Only 2 to 5 % of the protein in the small intestine escapes digestion and absorption. Some of the ingested protein that enter colon is digested by bacterial action. The protein in the stool is not of dietary origin but comes from bacteria and cellular debris.

**12. Lingual lipase is secreted by**

A. Parotid glands
B. Mandibular glands
C. Ebner's glands
D. Gastric glands

**Ans. 12. (C)**

Lingual lipase is secreted by Ebner's glands on the dorsal surface of the tongue. It catalyses hydrolysis of triglycerides to fatty acids and monoglycerides acylated in 2-position.

**13. Which of the following statements regarding 'Micelles' is NOT TRUE**

A. They are formed by lipids and bile salts
B. They solubilizes lipids
C. They transport lipids to enterocytes
D. They contain fatty acids, monoglycerides, and cholesterol in their hydrophilic part

**Ans. 13. (D)**

Micelles are macromolecules formed by bile salts. When the concentration of bile salts in the intestine is high lipids and bile salts interact spontaneously to form micelles. They contain fatty acids, monoglycerides, and cholesterol in their hydrophobic centers. Micellar formation further solubilizes the lipids and provides a mechanism for their transport to the enterocytes.

**14. Vitamin $B_{12}$ is absorbed in**

A. Ileum
B. Upper part of duodenum
C. Jejunum
D. Colon

**Ans. 14. (A)**

Vitamin $B_{12}$ binds to intrinsic factor, a protein secreted by the stomach, and the complex is absorbed across the ileal mucosa.

**15. Transferrin is**

A. Iron transport protein
B. Stored form of iron
C. Heme breakdown product
D. None of the above

**Ans. 15. (A)**

Transferrin is an iron transport protein. In the plasma $Fe^{2+}$ is converted to $Fe^{3+}$ and bound to transferrin which has two iron-binding sites. Normally, transferrin is about 35% saturated with iron.

# 14

# Cerebrospinal Fluid

1. **Subarachnoid space is present between**
   - **A.** Dura mater and arachnoid mater
   - **B.** Dura mater a pia mater
   - **C.** Arachnoid mater and pia mater
   - **D.** Between the layers of dura mater

**Ans. 1. (C)**

2. **Present between two layers of dura mater is**
   - **A.** Superior sagittal venous sinus
   - **B.** Inferior sagittal venous sinus
   - **C.** Both A and B
   - **D.** None of the above

**Ans. 2. (C)**

3. **Cerebrospinal fluid is secreted by**
   - **A.** Choroid plexuses of lateral ventricles
   - **B.** Choroid plexuses of lateral ventricles and third ventricle
   - **C.** Arachnoid villi
   - **D.** Choroid plexuses of third ventricle

**Ans. 3. (B)**

Choroid plexuses are the tufts of capillaries that project into the cavities of ventricles. CSF is secreted by Choroid plexuses, mainly the large plexuses of lateral ventricles but to some extent by the choroids plexuses of third ventricle. Additional amounts are secreted by the ependymal surfaces of the ventricles and arachnoidal membrane, and small amount comes from the brain itself through the perivascular spaces that surround the blood vessels entering the brain.

4. **Hydrocephalus is characterized by the following, EXCEPT**
   - **A.** Obstruction to the outflow of CSF
   - **B.** Increased intracranial tension

C. Excess of CSF in ventricles or in subarachnoid space
D. Hypertrophy of brain tissue

**Ans. 4. (D)**

Hydrocephalus means pathological accumulation of CSF in the cranial vault. It is due the obstruction to the outflow of fluid. Therefore, there is excess CSF in the ventricular system or in the subarachnoid space depending on the site of obstruction. Due to obstruction, CSF accumulates in the ventricles causing enlargement in their size and rise in intracranial tension. In this condition brain matter is pressurized and considerable damage to brain occurs.

**5. The path of circulation of CSF is**

A. Lateral ventricles → Third ventricle → Aqueductus Sylvii → Fourth ventricle→ Subarachnoid space
B. Lateral ventricles → Fourth ventricle → Aqueductus Sylvii → Third ventricle → Subarachnoid space
C. Lateral ventricles → Third ventricle → Fourth ventricle →Aqueductus Sylvii → Subarachnoid space
D. Lateral ventricles → Subarachnoid space → Third ventricle → Aqueductus Sylvii → Fourth ventricle

**Ans. 5. (C)**

**6. Absorption of CSF Occurs**

A. By the pia mater
B. Via arachnoid villi into the dural sinuses and spinal veins
C. Via arachnoid villi into the internal carotid artery
D. By choroid plexuses

**Ans. 6. (B)**

CSF is absorbed mainly via the arachnoid villi into the dural sinuses and the spinal veins; to a minor degree it may pass along the sheaths of the cranial nerves into the cervical lymphatics and also into the perivascular spaces.

# 15

# Male Reproductive System

**1. Cowper's glands are also known as**

**A.** Bulbourethral glands

**B.** Prostatic glands

**C.** Seminal glands

**D.** Testicular glands

**Ans. 1. (A)**

**2. All of the following is true regarding spermatogenesis, EXCEPT**

**A.** It continues throughout the life

**B.** It occurs in seminiferous tubules

**C.** It occurs in seminal vesicles

**D.** Maturation of spermatids into spermatozoa occurs in Sertoli cells

**Ans. 2. (C)**

Seminal vesicles are secretory glands. Each seminal vesicle is lined by the secretory epithelium that secretes seminal fluid. Spermatogenesis occurs in the seminiferous tubules and sperms are derived from the germinal cells of the seminiferous tubules. Maturation of spermatids into spermatozoa occurs in the deep folds of the cytoplasm of the Sertoli cells. Spermatogenesis continues throughout the life.

**3. Total number of type B spermatogonia produced after mitotic divisions of each primordial spermatogonium are**

**A.** 12 **B.** 16

**C.** 9 **D.** 20

**Ans. 3. (B)**

Primordial spermatogonium also known as type A spermatogonium undergoes 4 mitotic divisions to give rise to 16 type B spermatogonia.

od-testis barrier in human beings is derived n

A. Sertoli cells B. Leydig cells
C. Germinal cells D. Tunica albuginea

**Ans. 4. (A)**

The membranes of the Sertoli cells are tightly adherent to one another at their bases and sides, forming a permeability barrier that limits the transport of many substances like large protein molecules such as immunoglobulins, from blood to seminiferous tubular lumen. This barrier is known as blood-testis barrier. Leydig cells are present between seminiferous tubules and they produce androgenic steroids. Germinal cells give rise to sperms. Tunica albuginea is a fibrous connective tissue covering the testis.

**5. Each spermatid carries**

A. 46 chromosomes
B. 23 chromosomes
C. 12 chromosomes and one sex chromosome
D. None of the above

**Ans. 5. (B)**

Each spermatid is a haploid cell having 23 chromosomes in it.

**6. Entire period of spermatogenesis from germinal cell to sperm takes about**

A. 64 days B. 24 days
C. 36 days D. 18 days

**Ans. 6. (A)**

**7. In fetal testis, H-Y antigen is secreted by**

A. Leydig cells
B. Primordial germ cells
C. Sertoli cells
D. Trophoblast cells

**Ans. 7. (C)**

At 7 weeks gestation, Sertoli cells begin to form and secrete H-Y antigen which is under control of the Y chromosome. H-Y antigen is responsible for induction of testicular organogenesis. Leydig cells secrete testosterone in response to chorionic gonadotropin secreted by the syncytiotrophoblastic cells of placenta. Trophoblast cells are responsible for the implantation of blastocyst in the uterus and they also give rise to fetal membranes.

**8. Following are the functions of Sertoli cells, EXCEPT**

**A.** They provide mechanical support to the maturing sperms
**B.** They secrete müllerian duct inhibiting factor, which causes regression of müllerian duct system
**C.** They synthesize estradiol from androgenic precursors
**D.** They secrete testosterone

**Ans. 8. (D)**

Sertoli cells secrete small amounts of estradiol. Testosterone is converted to estradiol by the action of aromatase enzyme which is a membrane bound enzyme present in the sertoli cells. Testosterone is secreted by the Leydig cells.

**9. Acrosome in the sperm is derived mainly from**

**A.** Endoplasmic reticulum
**B.** Nucleus
**C.** Golgi apparatus
**D.** Cytoplasmic vacuoles

**Ans. 9. (C)**

**10. Capacitation of spermatozoa occurs**

**A.** In the seminal vesicle
**B.** In the Sertoli cells
**C.** In the epididymis
**D.** After ejaculation of the spermatozoa into the uterus of the female

**Ans. 10. (D)**

Once ejaculated into the female, the spermatozoa move up the uterus to the isthmus of the fallopian tubes, where they slowdown and undergo capacitation. This maturation process involves two components: increasing the motility of the spermatozoa and facilitating their preparation for the acrosome reaction.

**11. Clotting enzyme in the semen is derived from**

**A.** Prostatic glands
**B.** Seminal vesicles

C. Urethral glands

D. Seminiferous tubules

**Ans. 11. (A)**

Clotting enzyme is derived from prostatic glands. It causes the fibrinogen of the seminal vesicle fluid to form a weak coagulum that holds the semen in the deeper regions of the vagina where the uterine cervix lies.

**12. Hormone controlling testosterone secretion by Leydig cells in adult testes is**

**A.** Growth hormone **B.** FSH

**C.** Luteinising hormone **D.** Estrogen

**Ans. 12. (C)**

Testosterone secretion is under the control of luteinising hormone. Follicle-stimulating hormone stimulates Sertoli cells, without this stimulation spermiogenesis will not occur. Growth hormone is necessary for controlling background metabolic functions of testes. It also promotes early divisions of spermatogonia. Estrogen is formed from the testosterone by the Sertoli cells when they are stimulated by FSH and estrogen is probably also essential for spermiogenesis.

**13. Cryptorchidism means**

**A.** Failure of testis to descend from abdomen into the scrotum

**B.** Failure of testes to produce sperms

**C.** Failure of growth of testes

**D.** Hyperfunction of testes

**Ans. 13. (A)**

Cryptorchidism means failure of testis to descend from abdomen into the scrotum. Testicular descend to the inguinal region depends on MIS (müllerian inhibiting substance) and descend from inguinal region to the scrotum depends on other factors.

**14. Main nutritional supply for the spermatozoa is**

**A.** Glucose **B.** Fructose

**C.** Lactose **D.** Galactose

**Ans. 14. (B)**

Main nutritional supply of for the spermatozoa is fructose in the seminal fluid secreted by the seminal vesicles.

**15. Androstenedione is converted to one of the following estrogens**

A. Estradiol B. Estrone
C. Estriol D. None of the above

**Ans. 15. (B)**

Circulating androstenedione secreted by the testis is converted to estrone by the action of aromatase secreted by the Sertoli cells. Many non-endocrine tissues (e.g. brain, skin, adipose tissue, liver) also have aromatase, which converts androgens to estrogens in men, for example, testosterone is converted to estradiol and androstenedione is converted to estrone.

**16. Dihydrotestosterone is formed from the testosterone by the action of**

A. Aromatase B. 20, 22-desmolase
C. 5α-reductase D. 17α-hydroxylase

**Ans. 16. (C)**

Dihydrotestosterone is formed from testosterone by the action of 5α-reductase in some target cells. 5α-reductase is of two types.

Type 1, 5α-reductase is present in skin throughout the body and is the dominant enzyme in the scalp. Type 2, 5α-reductase is present in the genital skin, prostate gland, and other genital tissues.

20, 22-desmolase is the rate limiting enzyme, which converts cholesterol to pregnenolone. 17α-hydroxylase present in the Leydig cells, which causes hydroxylation of pregnenolone at 17th position to form 17α-hydroxypregnenolone, which is then converted to dehydroepiandrosterone. Dehydroepiandrosterone and androstenedione are the precursors in the synthesis of testosterone.

**17. Azoospermia is defined as**

A. 20 million sperm/ml of ejaculate
B. 10-20 million sperm/ml of ejaculate
C. 5-10 million sperm/ml of ejaculate
D. 5 million sperm/ml of ejaculate

**Ans. 17. (D)**

**18. Following are the effects of testosterone, EXCEPT**

A. Differentiation of Wolffian duct system into the epididymis, vas deferens, and seminal vesicles
B. Stimulation of organogenesis of urogenital sinus and tubercle into prostate gland, penis, urethra and scrotum

C. Spermatogenesis
D. Growth of penis, seminal vesicles, musculature, skeleton and larynx

**Ans. 18. (B)**

Dihydrotestosterone stimulates organogenesis of urogenital sinus and tubercle into prostate gland, penis, urethra and scrotum.

**19. In the developing fetus, the stimulus for the testosterone synthesis is**

A. Human chorionic gonadotropin
B. Luteinising hormone
C. Growth hormone
D. FSH

**Ans. 19. (A)**

In the developing fetus testosterone synthesis is stimulated by human chorionic gonadotropin (hCG), which is a placental hormone. In adults, luteinising hormone controls testosterone synthesis by Leydig cells.

**20. Relaxin in man is produced in**

A. Seminal vesicles
B. Seminiferous tubules
C. Prostate gland
D. Corpus luteum in women

**Ans. 20. (C)**

Relaxin is produced in the prostate gland in males. It helps maintain sperm motility and aid in sperm penetration of ovum.

It is produced in corpus luteum, uterus, placenta and mammary glands in women.

# 16

# Female Reproductive System - I

**1. Second polar body is extruded from the oocyte**

**A.** At the first meiotic division
**B.** At menarche
**C.** At the end of the menstrual cycle
**D.** Few hours after sperm enters the oocyte

**Ans. 1. (D)**

Until the entry of sperm, secondary oocyte remains in the metaphase of the second meiotic division. A few hours after the sperm enters the oocyte, the nucleus divides again and a second polar body is expelled, thus forming the mature ovum, which contains 23 chromosomes.

**2. Fertilization of the ovum normally occurs in**

**A.** In the distal portion of the oviduct (Ampula)
**B.** Body of the uterus
**C.** Cervix of the uterus
**D.** At the junction of the oviduct and the body of the uterus

**Ans. 2. (A)**

**3. First meiotic division of the primary oocyte is completed**

**A.** Just prior to ovulation
**B.** After the entry of sperm in the oocyte
**C.** Just prior birth
**D.** At the end of menstrual cycle

**Ans. 3. (A)**

Primary oocytes enter a prolonged stage of prophase (diplotene stage) of first meiotic division and remain in this stage until ovulation first occurs. Shortly before ovulation, primary oocyte nucleus divides by meiosis and a first polar body is expelled from the nucleus of the oocyte. Thus first meiotic division is completed just before ovulation and now the primary oocyte becomes secondary oocyte.

**4. Until the entry of the sperm, second meiotic division of the secondary oocyte remains in**

**A.** Prophase **B.** Metaphase
**C.** Anaphase **D.** Telophase

**Ans. 4. (B)**

**5. Normally, the number of ova expelled from ovarian follicle during the monthly sexual cycle (menstrual cycle) is**

**A.** One **B.** Two
**C.** Four **D.** Five

**Ans. 5. (A)**

At the start of each monthly sexual cycle several of the primordial follicles enlarge but one of the follicles in one ovary starts to grow rapidly on about the sixth day and becomes a dominant follicle, while the others regress forming atretic follicles. Thus, in each monthly sexual cycle, only one ovum is expelled from the ovarian follicle.

**6. Oocyte-maturation inhibiting factor is secreted by**

**A.** Granulosa cells
**B.** Spindle cells
**C.** Theca cells
**D.** Granulosa cells and theca cells

**Ans. 6. (A)**

During the course of maturation, the primary oocyte first becomes surrounded by basal lamina to form a primordial follicle; the flat spindle cells inside the basal lamina become cuboidal cells and thus form primary follicle. These cuboidal cells multiply to form granulosa cells. Throughout the childhood, the granulosa cells are believed to secrete an oocyte-maturation inhibiting factor, which keeps the ovum in its primordial state, suspended during this entire time in the prophase stage of meiotic division.

**7. At the time of birth the oocyte is a**

**A.** Primary oocyte in the metaphase of first meiotic division
**B.** Primary oocyte in the prophase of first meiotic division

C. Primary oocyte in the anaphase of first meiotic division
D. Secondary oocyte in the metaphase of second meiotic division

**Ans. 7. (B)**

**8. Stromal cells outside the basal lamina in the maturing follicle differentiates to form**

A. Granulose cells  B. Theca cells
C. Trophoblast cells  D. Both A and B

**Ans. 8. (B)**

In maturing follicles, the stromal cells outside the basal lamina differentiate to form theca cells. The layer just outside the basal lamina is called theca interna and external to theca interna lies theca externa. Theca interna cells produce androstenedione and testosterone, which diffuse into granulosa cells. Androstenedione and testosterone are converted to estrone and estradiol respectively.

**9. Follicular fluid secreted by granulose cells contains high concentration of**

A. Progesterone  B. Testosterone
C. Estrogen  D. LH

**Ans. 9. (C)**

**10. Early growth of primary follicle up to the antral stage is stimulated mainly by**

A. FSH  B. LH
C. hCG  D. Estrogen

**Ans. 10. (A)**

During preovulatory phase, the dominant gonadotropic hormone is FSH. The early growth of the primary follicle up to the antral stage is stimulated mainly by FSH alone. Thereafter accelerated growth of follicle is stimulated by combined action of estrogen, FSH and LH.

**11. The effects of estrogen does not include**

A. Inhibition of myometrial contractions
B. Growth of ovarian follicles
C. Salt and water retention
D. Promotion of the closure of epiphysial plates in the long bones

**Ans. 11. (A)**

Estrogens increase the amount of contractile proteins, actin and myosin, in the myometrium and, thereby, increase spontaneous muscular contractions. Estrogens also sensitize the myometrium to the action of oxytocin, which promotes uterine contractility.

**12. Theca interna cells of ovarian follicle produce**

A. Androstenedione B. Estriol
C. Estradiol D. Estrone

**Ans. 12. (A)**

**13. The effects of progesterone does not include**

A. Growth of ovarian follicles
B. Inhibition of myometrial contractions
C. Stimulation of development of lobules and alveoli in the breast
D. Stimulation of secretory activity of endometrium

**Ans. 13. (A)**

Growth of ovarian follicles is stimulated by estrogen.

**14. Accelerated growth of follicles after antral stage leading to formation of vesicular follicles is caused by**

A. Estrogen
B. FSH
C. LH
D. Estrogen, FSH, and LH

**Ans. 14. (D)**

Accelerated growth of follicles after antral stage is caused by

a) Estrogen makes granulose cells more sensitive than ever to FSH by causing granulosa cells to form FSH receptors.

b) FSH and estrogens combine to promote LH receptors on the granulosa cells, thus causing LH stimulation of these cells in addition to FSH stimulation and cause rapid increase in follicular secretion.

c) Increasing estrogen and LH act together to cause proliferation of follicular thecal cells and increase their secretion as well.

Therefore, once the antral follicle begins to grow their further growth occurs rapidly.

**15. Without preovulatory surge of this hormone ovulation will not occur**

A. FSH B. LH
C. Estrogen D. Progesterone

**Ans. 15. (B)**

LH secretion is under the control of hypothalamic hormone, GnRH
(Gonadotropin-releasing hormone). GnRH is normally secreted in episodic bursts, and these bursts produce the circhoral peaks of LH secretion. Frequency of the GnRH is increased by estrogens and decreased by progesterone. The frequency increases late in the follicular phase of the ovarian cycle, culminating in the LH surge. At the time of midcycle LH surge the sensitivity of the gonadotropes (LH and FSH) is greatly increased because of their exposure to GnRH pulses of the frequency that exist at this time. This self-priming effect of GnRH is important in producing maximum LH response.

**16. Inhibin hormone is secreted by**

**A.** Sertoli cells in men
**B.** Sertoli cells in men and granulose cells in women
**C.** Leydig cells in men
**D.** Leydig cells in men and theca cells

**Ans. 16. (B)**

Inhibin hormone is secreted by Sertoli cells in men and granulosa cells in women. Two types exist, namely, Inhibin A and inhibin B. It is inhibin B that is FSH regulating inhibin in adult men and women. It inhibits FSH secretion.

**17. Corpus luteum of the pregnancy secretes**

**A.** FSH, estrogen, pogesterone
**B.** LH, estrogen, progesterone
**C.** Estrogen, progesterone
**D.** Estrogen, progesterone and relaxin

**Ans. 17. (D)**

Corpus luteum of the pregnancy secretes estrogen, progesterone and relaxin. Relaxin helps maintain pregnancy by inhibiting myometrial contractions.

**18. Functional life span of corpus luteum is extended by**

**A.** FSH and LH **B.** LH
**C.** LH and hCG **D.** FSH

**Ans. 18. (C)**

Anterior pituitary hormone, LH prolongs the life of corpus luteum usually to about 12 days. Placental hormone also has similar effect of prolonging the life of corpus luteum, usually maintaining it for atleast the first 2 to 4 months of pregnancy.

**19. The hormone secreted in the greatest amount by the corpus luteum in the postovulatory phase is,**

A. Progesterone
B. Estradiol
C. Estrone
D. Inhibin

**Ans. 19. (A)**

Both FSH and LH levels fall after their midcycle peak but remain sufficiently high to stimulate the lutein cells to secrete estradiol, estrone, progesterone and inhibin. But the hormone secreted in the greatest amount by the corpus luteum in this phase is progesterone. In fact, plasma concentrations of progesterone during the entire ovarian cycle are higher than those of estradiol or estrone. The units of concentration for progesterone (ng/ml) are 1000 times greater than for estrogen (pg/ml).

**20. Inhibin secreted by the corpus luteum inhibits secretion of**

A. Estrogen
B. Progesterone
C. LH
D. FSH

**Ans. 20. (D)**

**21. During the preovulatory stage of ovarian cycle, the dominant steroid that is secreted by the ovary is**

A. Estradiol
B. Estriol
C. Estrone
D. None of the above

**Ans. 21. (A)**

During the preovulatory stage of ovarian cycle, the dominant steroid that is secreted by the ovary is Estradiol mainly under the influence of FSH, which is a dominant gonadotropic hormone during this phase.

**22. Ovarian cycle is characterized by the following EXCEPT**

A. Plasma concentrations of progesterone during entire cycle are higher than that of estradiol
B. Progesterone peak occurs at about 6 days after ovulation which coincides with the mid-luteal peak of estradiol
C. FSH surge two days before ovulation triggers ovulation

**D.** Both FSH and LH levels fall after their midcycle peaks

**Ans. 22. (C)**

There is a surge in LH secretion that triggers ovulation, and ovulation normally occurs about 9 hours after the peak of LH surge at midcycle.

**23. Circulation of progesterone occurs by binding with**

**A.** Gonadal steroid – binding globulin (GBG)
**B.** Albumin
**C.** Corticosteroid – binding globulin (CBG)
**D.** Both albumin and corticosteroid – binding globulin (CBG)

**Ans. 23. (D)**

About 2% of the circulating progesterone is free, 80% is bound to albumin and 18% is bound to corticosteroid – binding globulin.

About 2% of the circulating estradiol is free, 60% is bound to albumin and 38% is bound to gonadal steroid – binding globulin.

**24. Oligomenorrhea refers to**

**A.** Failure to have menstrual periods
**B.** Bleeding from the uterus between menstrual periods
**C.** Reduced frequency of menstrual periods
**D.** Excessive loss of blood during menstrual periods

**Ans. 24. (C)**

Failure to have menstrual periods is called amenorrhea. Bleeding from the uterus between menstrual periods is called metrorrhagia. Reduced frequency of menstrual periods is called oligomenorrhea. Excessive loss of blood during menstrual periods is called menorrhagia.

17

# Female Reproductive System - II Pregnancy and Lactation

**1. In human placenta, 16-hydroxydehydroepiandrosterone is converted to**

A. Estriol  B. Estradiol
C. Estrone  D. Testosterone

**Ans. 1, (A)**

Fetal ACTH stimulates the secretion of DHEA (Dehydroepiandrosterone) and DHEAS (Dehydroepiandrosterone sulfate) from the fetal zone of fetal adrenal cortex. Some of the DHEA and DHEAS are taken up by the fetal liver and converted to 16-hydroxy- dehydroepiandrosterone or 16-hydroxy-dehydroepiandrosterone sulfate and released into the fetal blood. These steroids can also be taken up by the placenta and converted to estriol.

**2. Relaxin is produced in the**

A. Corpus luteum
B. Placenta
C. Both (A) and (B)
D. Posterior pituitary gland

**Ans. 2. (C)**

Relaxin is polypeptide hormone that is produced in the corpus luteum, uterus, placenta and the mammary glands in women and in the prostate gland in men.

**3. Placenta has following components**

A. Myometrium
B. Cytotrophoblast
C. Syncytiotrophoblast
D. Both Cytotrophoblast and Syncytiotrophoblast

**Ans. 3. (D)**

The placenta is formed by the tissues derived from the

embryo. The trophectoderm of blastocysts forms the placenta by invading the endometrial layer of uterus. The trophoblast forms two cell layers,

The inner layer forms the cytotrophoblast and inner layer forms the syncytiotrophoblast.

**4. Human placenta secretes following hormones, EXCEPT**

**A.** Progesterone
**B.** Estrogen
**C.** Human chorionic gonadotropin
**D.** Oxytocin

**Ans. 4. (D)**

Oxytocin is secreted by the posterior pituitary gland. Syncytial trophoblast cells of placenta secrete human chorionic gonadotropin progesterone and estrogen.

**5. Physiological anemia of pregnancy is due to**

**A.** Decrease in plasma iron binding capacity
**B.** Hypoactivity of bone marrow
**C.** Disproportionate increase in plasma volume
**D.** Deficiency of vitamin $B_{12}$

**Ans. 5. (C)**

Physiological anemia of pregnancy is due to Disproportionate increase in plasma volume. The increase in the plasma is not matched by an equivalent increase in red cell mass, therefore maternal hematocrit usually decreases resulting in physiological anemia of pregnancy.

**6. During labour, oxytocin increases uterine contractions by stimulating formation of**

**A.** $PGE_2$ **B.** $PGI_2$
**C.** $PGF_{2á}$ **D.** Nitric oxide

**Ans. 6. (C)**

During labour, oxytocin increases uterine contractions by two ways, (a) It acts directly on uterine smooth muscle cells to make them contract and (b) It stimulates formation of prostaglandin $F_{2á}$ ($PGF_{2á}$) in decidual cells. $PGF_{2á}$ is believed to stimulate or augment the contraction of myometrial cells.

**7. Principal estrogen secreted by human placenta**

**A.** Estriol **B.** Estradiol
**C.** Estrone **D.** None of the above

**Ans. 7. (A)**

Quantitatively, estriol is the major estrogen of human pregnancy. It is synthesized in placental trophoblasts. Estriol is produced primarily from the androgenic precursors formed in the fetal zone of the fetal adrenal cortex and the liver. Androgenic precursors, DHEAS (Dehydroepiandrosterone sulfate) forms estradiol and 16-OHDHEAS (16-hydroxydehydroepiandrosterone sulfate) forms estriol.

**8. In postmenopausal women, the dominant plasma estrogen is**

**A.** Estradiol **B.** Estrone
**C.** Estriol **D.** None of the above

**Ans. 8. (B)**

In postmenopausal women, the dominant plasma estrogen is estrone. Estrone is derived from the conversion of adrenocortical androstenedione in peripheral tissues mainly liver.

**9. Human placenta produces and secretes following hormones, except**

**A.** Human chorionic gonadotropin
**B.** Progesterone
**C.** Estrogen
**D.** Luteinising hormone (LH)

**Ans. 9. (D)**

Luteinising hormone is secreted by the gonadotropes, which are chromophilic secretory cells of the anterior pituitary gland.

**10. Placental hormone that is structurally and functionally similar to LH secreted by pituitary is**

**A.** Human chorionic gonadotropin (hCG)
**B.** Human placental lactogen (hPL)
**C.** Progesterone
**D.** Estrogen

**Ans. 10. (A)**

Like LH, hCG is also a luteotropic hormone, in that it maintains the function of corpus luteum until the feto-placental unit is autonomous in terms of hormone secretion.

**11. Human fetus is protected by immunosuppressive action of**

A. Human chorionic gonadotropin
B. Progesterone
C. Relaxin
D. Human placental lactogen

**Ans. 11. (B)**

Human fetus is protected by immunosuppressive action of progesterone. It may inhibit local host-graft rejection of the embryo through its effects as an immune suppressant.

**12. Postpartum milk production is stimulated by**

A. Growth hormone
B. Progesterone
C. Prolactin
D. Human chorionic gonadotropin

**Ans. 12. (C)**

Maternal milk production in postpartum period is stimulated by prolactin. Prolactin induces differentiation of presecretory cells into active secretory cells and increases the synthesis of fatty acids, phospholipids, and milk protein, casein, lactalbumin and α-lactoglobulin. The enzymes necessary for the synthesis of lactose are also induced by prolactin.

**13. In pregnancy, estrogen causes**

A. Proliferation of decidual layers of uterus
B. Increase in uterine blood flow
C. Both (A) and (B)
D. None of the above

**Ans. 13. (C)**

**14. During pregnancy, development of lobuloalveolar structure in the breast is caused by**

A. Progesterone and estrogen
B. Only progesterone
C. Only estrogen
D. Human chorionic gonadotropin

**Ans. 14. (A)**

**15. Uterine quiescence until the end of the pregnancy is achieved by the action of**

A. Estrogen
B. Prolactin
C. Oxytocin
D. Progesterone

**Ans. 15. (D)**

Uterine quiescence is thought to be achieved by the action of progesterone on the myometrium to suppress prostaglandin synthesis and inhibit oxytocin receptor expres-sion. Progesterone reduces the excitability of smooth muscle cells as a result of hyperpolarisation, and to prevent gap junction formation, resulting in reduced communication between smooth muscle cells.

**16. Uterine contractions is stimulated by**

A. Estrogen B. Oxytocin
C. Prostaglandins D. All of the above

**Ans. 16. (D)**

**17. Respiratory changes during pregnancy include**

A. Increased respiratory rate
B. Increased tidal volume
C. Respiratory alkolosis
D. All of the above

**Ans. 17. (D)**

During pregnancy, tidal volume is increased and there is slight increase in respiratory rate. This may be due to the effect of progesterone, which appears to increase the sensitivity of respiratory centers to $PCO_2$ levels in arterial blood. The increased ventilation tends to cause decreased arterial $PCO_2$ thus, pregnant women frequently have respiratory alkolosis and compensatory metabolic acidosis.

**18. Increased thyroxine production during pregnancy is because of**

A. Human chorionic gonadotropin
B. Human chorionic thyrotropin
C. Both human chorionic gonadotropin and human chorionic thyrotropin
D. Progesterone

**Ans. 18. (C)**

Increased thyroxine production during pregnancy is because of human chorionic gonadotropin and human chorionic thyrotropin secreted by the placenta.

**19. Increased in plasma volume during pregnancy is due to**

A. Progesterone
B. Estrogen and progesterone induced release of rennin from kidneys

C. Increased cortisol secretion

D. All of the above

**Ans. 19. (D)**

Estrogen increases the production of angiotensinogen in the liver during pregnancy. Progesterone and estrogen increase the release of rennin from kidneys. These effects cause increase in plasma concentrations of angiotensin II. Aldosterone concentrations are therefore increased, which causes increase plasma volume. ACTH and cortisol concentrations in plasma are also increased during pregnancy and may contribute increased plasma volume; in addition progesterone itself may increase plasma volume.

**20. Eclampsia associated with pregnancy is characterized by following, EXCEPT**

A. Rise in arterial blood pressure

B. Greatly increased kidney output

C. Clonic seizures

D. Extreme vascular spasms

**Ans. 20. (B)**

In eclampsia there is extreme vascular spasm throughout the body, especially in kidneys as a result of which the kidney output is greatly reduced.

# 18

# Fetal and Neonatal Physiology

**1. In early pregnancy, fetal aldosterone and cortisol is synthesized from**
- **A.** Maternal estrogen
- **B.** Placental progesterone
- **C.** Fetal testosterone
- **D.** Placental human chorionic gonadotropin

**Ans. 1. (B)**

In early pregnancy, the fetus requires placental progesterone to synthesize aldosterone and cortisol, because fetal zone of the fetal adrenal cortex lacks the enzyme 3β–hydroxysteroid dehydrogenase. This enzyme is required for the conversion of pregnenolone to progesterone and therefore also required for the synthesis of glucocorticoids and mineralocorticoids. Beyond 10 weeks gestation, the fetal androgen no longer depends on placental progesterone for synthesis of aldosterone and cortisol.

**2. Viability of fetus can be judged by the levels of one of the following hormone in the maternal plasma**
- **A.** Estrone
- **B.** Estradiol
- **C.** Estriol
- **D.** ACTH

**Ans. 2. (C)**

Because fetoplacental unit requires a functioning fetal pituitary, fetal adrenal and placenta, the circulating levels of the plasma estriol of pregnant woman is used to demonstrate the viability of fetus.

**3. Fetal adrenal cortex secretes**
- **A.** Corticosterone
- **B.** Progesterone
- **C.** Dehydroepiandrosterone (DHEA)
- **D.** ACTH

**Ans. 3. (C)**

The chief product of fetal zone of adrenal cortex is dehydroepiandrosterone (DHEA), which is secreted as inactive sulfate ester. DHEA sulfate is converted in the fetal liver to 16α-hydroxydehydroepiandrosterone sulfate, which is aromatized in the placenta to form estriol (estriol is synthesized only by placenta and adult liver). After first trimester the inner zone of fetal adrenal cortex can secrete cortisol and aldosterone. Progesterone and corticosterone are synthesized, but not secreted, by the fetal adrenal cortex. ACTH is secreted by fetal pituitary gland.

**4. Within fetal circulation, the blood from the umbilical vein is carried to the inferior vena cava by**

**A.** Ductus umbilicus **B.** Ductus arteriosus
**C.** Foramen ovale **D.** Ductus venosus

**Ans. 4. (D)**

In fetal circulation the blood from the umbilical vein passes through the ductus venosus into the inferior vena cava bypassing liver.

**5. Foramen ovale is present between**

**A.** Right atrium and left atrium
**B.** Right ventricle and left ventricle
**C.** Pulmonary artery and descending aorta
**D.** Portal vein and inferior vena cava

**Ans. 5. (A)**

Foramen ovale is present between two atria, which allows blood to pass from the right atrium to left atrium. In fetal heart the hydrostatic pressure of the blood in the right atrium is greater than that in left atrium.

**6. For fetal energy supply, the substrate that is mainly used is,**

**A.** Triglycerides **B.** Glucose
**C.** Proteins **D.** Glycerol

**Ans. 6. (B)**

Fetus uses mainly glucose for energy. Glucose is supplied to the fetus primarily through transplacental transport.

**7. Fetal kidneys become capable of excreting urine during**

**A.** First three months of pregnancy
**B.** Last three months of pregnancy

C. Last ten days of pregnancy
D. Later half of pregnancy

**Ans. 7. (D)**

Fetal kidneys become capable of excreting urine in the later half of the pregnancy and urination occurs normally *in utero*. However, renal control systems for regulation of ECF, electrolyte balance and acid-base balance are almost non-existent until after midfetal life. And reach full development only after few months after birth.

**8. For formation of red cells fetus needs**

A. Vitamin $B_{12}$ and folic acid
B. Vitamin D
C. Vitamin C
D. All of the above

**Ans. 8. (A)**

Active form of folic acid is Tetrahydrofolate (THF). DNA synthesis requires methylation of Uridine to Thymidine. Serine provides the methyl group required. THF is requi-red to transfer this methyl group from Serine to Uridine in 2 steps:

a) Serine transfers methylene group to THF to form Glycine and Methylene – THF.

b) Methylene – THF transfers the methyl group to Uridine, and in the process, gets converted to Dihydrofolate (DHF).

**Methyl trap:**

Some of the methylene-THF formed gets converted into methyl-THF. The ability of the cell to use methyl-THF is related to its ability to keep or "trap" the folate inside the cell. (Methyl trap). Vitamin $B_{12}$ is required for the deme-thylation (removal of methyl group) of folate that has been taken up by the cell, allowing it then to be conjugated (that is, supplied with a polyglutamate tail), a reaction that prevents the folate from leaking out of the cell. Megaloblastic anemia occurs because the folate that gets trapped in the methyl trap is not recovered resulting in folate deficiency, which affects DNA synthesis in the cell.

**9. Formation of blood begins in**

A. Fifth week of intrauterine life
B. Third week of intrauterine life

**C.** Three months of intrauterine life
**D.** Seven days after fertilization of ovum

**Ans. 9. (B)**

Formation of blood starts in the 3rd week of intrauterine life. Between 3rd week to 3rd month of intrauterine life, erythropoiesis occurs in the mesoderm of the yolk sac.

**10. The most abundant surfactant protein in the fetal lungs is**

**A.** SP-A **B.** SP-B
**C.** SP-C **D.** SP-D

**Ans. 10. (A)**

Major surfactant proteins are SP-A, SP-B, SP-C, and SP-D. The most abundant of them is surfactant protein, SP-A. SP-A binds to lipids, calcium and sugars and works with the other component of surfactant to form an air interface with low surface tension.

**11. Women at risk of for premature labour are treated with one of the following steroids**

**A.** Dexamethasone
**B.** Estrone
**C.** Estradiol
**D.** Progesterone

**Ans. 11. (A)**

Glucocorticoids accelerate the terminal maturation of fetal lung, including stimulation of the endogenous production of surfactant proteins and stimulation of the incorporation of choline into phosphatidyl choline. Therefore women at risk of premature labour are treated with the synthetic glucocorticoids, for example, dexamethasone or beta-methasone. This reduces the risk of perinatal death secondary to respiratory distress.

**12. Major secretory product of fetal zone of fetal adrenal**

**A.** Corticosterone **B.** Cortisol
**C.** Dehydroepiandrosterone **D.** Aldosterone

**Ans. 12. (C)**

# READER SUGGESTIONS SHEET

*Please help us to improve the quality of our publications by completing and returning this sheet to us.*

Author/Title: **MCQs in Physiology** ***By Jayant J Makwana, Surendra S Wadikar***

Your name and address:

Phone and Fax:

e-mail address:

How did you hear about this book? [please tick appropriate box (es)]

☐ Direct mail from publisher ☐ Conference ☐ Bookshop

☐ Book review ☐ Lecturer recommendation ☐ Friends

☐ Other (please specify) ☐ Website

**Type of purchase:** ☐ Direct purchase ☐ Bookshop ☐ Friends

Do you have any brief comments on the book?

**Please return this sheet to the name and address given below.**


**JAYPEE BROTHERS**

**MEDICAL PUBLISHERS (P) LTD**

EMCA House, 23/23B Ansari Road, Daryaganj

New Delhi 110 002, India